Blackbirds of the Americas

Yellow-winged blackbirds

Austral blackbirds

Blackbirds of the Americas

Gordon H. Orians

With drawings by **Tony Angell**

Chopi

University of Washington Press   *Seattle and London*

Composition by University of Washington Department of
Printing, Seattle
Printing and binding by Toppan Printing Company, Tokyo.
Designed by Audrey Meyer

Library of Congress Cataloging in Publication data

Orians, Gordon H.
   Blackbirds of the Americas.

   Bibliography: p.
   Includes index.
   1. Blackbirds.   2. Birds—America.   I. Angell, Tony   II. Title
QL696.P24750737   1985   598.8′81   85-40352
ISBN 0-295-96253-4

Printed and bound in Japan

Yellow-winged blackbirds

Contents

Epaulet oriole

Preface

Bird watching has been a source of delight for people as long as recorded history informs us. The scientific study of birds has its origins deeply rooted in the joy of observing birds as they carry out the activities necessary for them to survive and leave offspring. Ornithology, like all other sciences, has become more abstract and obscure as it developed the theoretical superstructure appropriate to a mature science of the twentieth century. Nonetheless, most ornithologists never lose their sense of wonder while watching birds. Indeed, it is the fun of ornithology that motivates much of what is done and described in rather formal language in the scientific literature. Yet, if scientific ornithology is to provide society at large with the full range of its benefits, scientists must write, at times, for more general audiences so those whose appreciation of birds is avocational can enrich their joy and wonderment with deeper insights into the causes of bird behavior. This book is a contribution to that effort. We regard ourselves as following the lead of David Lack, former director of the Edward Grey Institute of Field Ornithology at Oxford University, England, whose books *The Life of the Robin, Enjoying Ornithology,* and *Swifts in a Tower* did so much to stimulate a flow of information and ideas between scientists and laypersons. It is, therefore, appropriate that the final push that enabled Orians to complete a draft of the manuscript was accomplished at the Edward Grey Institute.

This book is the product of a large dose of specialization, our concentration on the members of a particular family of birds, the American blackbirds or Icteridae. To specialize or not to specialize is one of the big questions. Since we have obviously done it, it is appropriate to ask, "What, if anything, have we gained as a result?" The main benefit from specialization is achievement of a depth of understanding about a problem that cannot be gained by examinations where breadth is the prime objective. Specialization is especially important for biologists because evolutionary pathways, although guided by processes believed to operate on all living organisms, are richly varied. Many of the questions we will consider in this book gain additional meaning because we are able to compare the behavior of a number of species of blackbirds in order to reveal both similarities and differences among them.

This book is also the product of combining two rather different specializations, those of the artist and the scientist. The values deriving from specialization are similar in science and the creative arts, presumably because of deep commonalities in what constitutes human creativity in all its forms. We hope that our collaboration will produce exactly this result as we describe, in words and pictures, the ways blackbirds find food, select places in which to breed, choose associates, participate in breeding, and communicate.

During the decade this book was being formed Tony Angell composed and illustrated a book on owls, one on crows and their relatives, and another on marine birds and mammals. Each of these books has to some degree been influenced and stimulated by the ongoing project of delving into the lives of the American blackbirds. Icterids can be subdued or raucous, colorful or drab, and simple or complex in their social systems. It has been a delightful challenge to discover ways to convey these ranges of attributes in illustrations.

The illustrations, then, seek to complement and extend what the words convey. Rather than being identification images or portraits, the drawings are composed and designed to provide information while also seeking to touch the esthetic sense of the reader. These goals may be of equal importance, for while a picture should transcend words, beauty can also stimulate the observer to look and think more deeply about a subject than words alone.

A work such as this one reflects contributions of many people. We are especially indebted to R. Haven Wiley and Christopher C. Perrins for reading the entire manuscript in draft and to Lynn Erckmann for help with the field work, production of a number of fine graphs, and general support in a variety of ways. Clean copy was produced with remarkable speed and accuracy from sometimes rather messy text by Barbara Peterson who, through her own extensive knowledge of the birds, was able also to make valuable suggestions. Loans of specimens were made by the Museum of Vertebrate Zoology, University of California, Berkeley, the U.S. National Museum in Washington, D.C., and the American Museum of Natural History in New York. The inspiration of the late David Lack, who devoted so much of his considerable talents to introducing a wide variety of people to the excitement of the avian world, has been with us repeatedly during our efforts to carry on the tradition he so ably served.

Gordon H. Orians
Tony Angell

Introduction

Blackbirds were among the first birds I encountered when I took up bird watching as a boy of seven. I remember red-winged blackbirds most vividly, partly because the brilliant plumages of the males, but mostly because their return to southern Wisconsin, where I lived, from more hospitable southern wintering grounds was the first certain sign that spring had arrived. Before I was ten I discovered grackles, cowbirds, and yellow-headed blackbirds.

My appreciation of blackbirds of all colors grew as my life became more and more intertwined with theirs. As an undergraduate at the University of Wisconsin I worked in the Zoology Museum where the curator, Robert Nero, was engaged in a detailed study of the breeding behavior of redwings. I gave them more than a casual look as I prepared specimens for future study. My academic advisor, Professor John Emlen, had worked on blackbirds for many years. His enthusiasm for those birds undoubtedly influenced my decision to carry out a comparative study of redwings and tricolored blackbirds for my doctoral thesis under Professor Frank Pitelka at the University of California at Berkeley. Having invested that much time and effort on blackbirds, I naturally decided to continue studying them when I joined the faculty of the University of Washington in Seattle, in August of 1960.

For more than two decades, with breaks now and again to study such diverse organisms as tent caterpillars and slugs, my favorite herbivores, and desert plants, I have tried to solve the puzzles of how these birds conduct their lives and why each species has its own particular behavior patterns, social organization, and ecology. From Canada to Argentina I have stalked blackbirds—black, yellow, red, and brown.

I have stayed with blackbirds because I got hooked on them. Spring is not complete without the raucous songs of territorial male yellow-headed blackbirds, so terrible by any normal standards of song quality, yet so deeply filed in my brain that the harshness actually sounds pleasant. But there is more to it than that. Blackbirds are common and easy to study, and they have virtually all of the forms of social organization known among birds. Closely related species that differ in ecology, behavior, and social organization are especially valuable as objects of study because, in the not too distant past, they shared a common ancestor and, hence, evolved their differences by diverging from a common base. Therefore, they reveal more clearly than would

otherwise be the case how environmental conditions influence ecological and behavioral adaptations.

To survive in an environment an organism requires food and shelter plus conditions that allow reproduction to take place. The kinds of resources available and how they are distributed in the environment strongly affect the value of different behaviors and social organization patterns. Therefore, I begin my discussion of blackbirds with foods, foraging, and nest sites, and then turn to the way blackbird societies are organized to exploit these resources. The functioning of these societies depends, in turn, on communication among their members. Birds, like people, communicate primarily by means of visual and vocal signals that are easy for us to perceive, although, as we shall see, they are often hard to understand. Our own communication signals are notoriously difficult to interpret and some of the causes of the problems are common to both people and blackbirds.

Blackbirds and Our Self-perceptions

By looking at other organisms we may develop new insights and perspectives about ourselves. We are naturally highly interested in our own structure, physiology, and behavior—so much so, in fact, that if all parts of a contemporary university dealing with the study of *Homo sapiens* were to be eliminated, not very much would remain. We have even made religious dogma out of this strong self-interest. The proper study of people is people—so we are told. Nevertheless, recognition of our place as only one of the millions of species that have evolved by the same natural processes that governed our own evolution, and that have evolved ways of handling the same basic set of problems that concern us, can help place our existence in a broader perspective. As our similarities with other species and our uniqueness become more apparent, we can better understand both our own lives and the lives of other animals. Blackbirds are not models of human evolution, and we must avoid temptations to draw quick analogies. They can help us understand the richness of solutions to life's challenges, however, among which our own solutions are but a small subset.

But we dally. Rains are filling the marshes and blackbirds are gathering food. If we do not join them we may miss some of the lessons they have to share with us.

*1.1 Yellow-backed orioles. The sexes look alike in this bright
tropical species.*

1. Blackbirds and Black Birds

Ornithologists recognize ninety-four species of blackbirds in the family Icteridae (from the Greek, meaning "yellow" or "jaundice"!). They are restricted entirely, except for accidental stragglers, to the New World, but within our hemisphere they live almost everywhere. They breed from Alaska and northern Canada south to Tierra del Fuego. They range in size from the crow-sized oropendolas of tropical forests to sparrow-sized marsh and grassland species. They live in nearly every habitat except tundra, and their social systems include most of the full range found among birds. Their abundance and conspicuousness have made them attractive subjects for scientific investigation. Their success in exploiting habitats that have been modified by human activity has led many of them to increase in abundance since European settlement of the New World. It has also brought some of them into direct conflict with human interests as crop pests.

Not all birds ornithologists classify as blackbirds are in fact black. Meadowlarks, despite being close relatives of redwings, are predominantly yellow and brown, while orioles, revealing their blackbird ancestry with patches of black, nevertheless advertise themselves in brilliant yellows and oranges (fig. 1.1).

Nor are all black birds classified as blackbirds. The European blackbird, of which four and twenty were served to the king, is really a kind of thrush (fig. 1.2). Nonetheless, this species, one of the most common and conspicuous of European birds, is the source of the name given to the American blackbirds. Early English colonists in North America applied familiar or descriptive names to the new birds they found, without much concern for the relationships later recognized by ornithologists. Not only were blackbirds named after *the* blackbird, but orioles were named because of their resemblance to the golden oriole of Europe, a bird belonging to quite a different family. Meadowlarks reminded the settlers of the true larks of the Old World. Nothing in Europe resembled some of the tropical blackbirds and so caciques received a Spanish name for "chief."

Experts agree that the icterids are all closely related, having recently evolved from a small, finchlike ancestor (fig. 1.3), but there is lack of agreement as to details of relationships within the family. Most similar to the presumed finchy ancestors, but remarkable in their breeding adaptations, are the *cowbirds*. With the exception of one south temperate species that incubates its

1.2 The European blackbird, not a member of the blackbird family but a thrush.

9

1.3 *Evolutionary tree of the Icteridae. From a cowbirdlike ancestor, separate lines of evolution lead to the redwings, orioles, and meadowlarks, (left), grackles (above), and oropendolas and caciques (below).*

own eggs and raises its own young, the cowbirds are all brood parasites—like the famous cuckoo they lay their eggs in the nests of other species who incubate them and feed the young. Interestingly, several of the cowbirds use other blackbirds as principal hosts, while other cowbirds have a host list that includes most of the small to medium-sized birds that breed within their ranges.

The *grackles* are a group of primarily black, slender-billed, terrestrial blackbirds widely distributed in North, Central, and northern South America (fig. 1.4). Grackles differ from each other primarily in size and in the degree to which males are larger than females. They tend to favor wet habitats, and several species have adapted well to urban environments, finding well-watered lawns suitable marsh substitutes. In parts of tropical America some of the species are generally confined to towns and villages, being rarely encountered in more pristine environments.

Blackbirds are among the most common and conspicuous birds of marshes throughout the Americas. A North American marsh would not seem complete without its red-winged blackbirds. Residents of Argentina make equally strong associations between marshes and their yellow-winged and scarlet-headed blackbirds. The association between blackbirds and marshes has puzzled ornithologists for many decades because there are no obvious morphological adaptations of blackbirds to a marsh environment. They do not have webbed feet or a bill that would filter algae from the water. The solution to this puzzle, as we shall soon see, lies in more subtle adaptations of bills and feeding behavior. Marsh-nesting blackbirds are small to medium, males are blackish, females are variably colored, and they exhibit some of the more interesting social systems found in the family. They are also the best known of the blackbirds, probably among the best known of all bird species.

1.4 *The great-tailed (upper left), boat-tailed (upper right), and common grackle (bottom) are all found in North and Central America.*

The *orioles* are believed to be closely related to the marsh blackbirds, their differences reflecting adaptations to foraging in trees and shrubs in rather open environments. The center of oriole evolution was evidently Mexico, where the greatest number of species is found today. Although environments appear to be highly suitable for them there, fewer orioles occur in South America. A few of the South American orioles are primarily black but, in general, orioles are among the brightest-colored members of the family. They also include some of its finest songsters.

Grasslands are the homes of *meadowlarks* and their allies. Their streaked backs blend in well with their environments, making them very difficult to see from above when they squat on the ground. Standing upright, however, they can display their brightly colored breasts and bellies, yellow in the North American species, red in the South American ones. North American grasslands also house the only inter-hemispheric migrant in the family, the *bobolink*, which winters in the pampas of Argentina. Tropical grasslands with scattered trees, commonly known as savannas, are also home for a number of rather nondescript, mostly black blackbirds of uncertain relationships. They primarily forage on the ground but nest in trees and shrubs. Because they live in areas where few ornithologists have worked, they are among the least known blackbirds. Their social organization holds a number of surprises, however, some of which we will explore later.

Tropical forests and woodlands are the habitats of the spectacular *oropendolas* and *caciques*. The very large, hanging nests of oropendolas, concentrated in colonies in large, usually isolated trees, are among the most characteristic sights of the American tropics (fig. 1.5). The smaller caciques include some highly colonial species with bright color patches, as well as mostly black, solitary species that forage in dense forest understory. None of the members of this group occur north of central Mexico or south of northern Argentina, but within the tropics they are common and some are among the most conspicuous of all birds.

Listing all of the common and scientific names of the nearly one hundred icterids would intrude unnecessarily into the flow of ideas and information in this and subsequent chapters, but I provide a list of the common and scientific names of the species in Appendix A. In addition, existing information on plumage, habitat, foods and feeding, and social organization of blackbirds is summarized in Appendix B. Because information on many of these points is lacking for many tropical species, this account is really a progress report on our understanding of blackbirds and what they have taught us about avian adaptations. We can expect them to be instructive subjects for study for many years to come.

Geological History of South America

The origins and evolutions of blackbirds have been molded by the development of the continents of the New World. Continents have moved, sea levels have changed, glaciers have advanced and retreated over the landscape, and climatic changes have caused major shifts in the distributions of vegetation types. All these events have left their imprint upon blackbirds as we can see them today.

At the end of the Paleozoic era, about 225 million years ago, all of the continents were united together into a supercontinent, Pangaea. How Pangaea was formed is unknown but, geologically speaking, it did not last long. In fact, by the middle Triassic, about 220 million years ago, it had separated into a large northern continent, Laurasia, and a still larger southern one, Gondwanaland. During the late Cretaceous, about 135 million years ago, South America and Africa drifted apart and from then on, until the Pliocene epoch, about three million years ago, when the Central American isthmus was formed, South America was isolated from all the other continents by broad expanses of water. Most of the very distinctive faunal elements of South America evolved during this long period. Among the mammals, sloths, armadillos, anteaters, hystricomorph rodents (porcupines, guinea pigs, and their relatives), and marmosets and cebid monkeys are unlike mammals found elsewhere, and even stranger ones are known only as fossils. The ornithologically inclined visitor to South America finds woodcreepers, ovenbirds, antbirds, flycatchers, manakins, cotingas, plant cutters, hummingbirds, puffbirds, toucans, screamers, trumpeters, and sun bitterns, all very distinct from birds of the Old World. A few of these, such as the flycatchers and hummingbirds, have spread northward and are an important part of the current North American avifauna, but most of them have not penetrated so far. They are the birds that United States bird-watchers hope to glimpse on their visits to southern Texas and Arizona, or which are high on the list of things to see on that long-awaited Central or South American excursion.

Modern South America was shaped by the rise of the lofty Andes, which occurred quite recently in the upper Pliocene and lower Pleistocene epochs. This upward thrust buckled and folded the landscape into a spectacular series of ranges towering to more than five thousand meters, separated by deep desert basins, some of them without surface drainage. Great lakes formed in them during the glacial period. Moisture-bearing winds from the Atlantic are forced to rise as they encounter the Andes. As the air cools it dumps its water copiously on the flanks of the mountains. Dense forests clothe the eastern flanks of the Andes south to

1.5 A colony of Montezuma oropendolas occupying part of a large kapok tree in eastern Guatemala.

Bolivia, with patches occurring in favorable locations as far south as northern Argentina. The rise of the Andes was followed by the raising of the Amazon Basin above sea level. This huge, relatively flat area, covering approximately two million square kilometers, contains the largest block of tropical forest in the world. At some unknown time and place during these events a few finches began to augment their diet by exposing hidden prey by the gaping movements of their bills. They did not know that they would give rise to blackbirds.

Tropical Climates

During the past one hundred thousand years tropical America has been profoundly affected by climatic fluctuations. Glaciers did not spread over the land as they did in North America. There were expansions of montane glaciers in the southern Andes, but these had a trivial effect on the continent compared with the changes in rainfall and temperature that occurred in tropical and subtropical regions.

These hundred thousand years have witnessed at least three alternating episodes of wet and dry in tropical South America and Central America, the two latest dry periods dated at about eleven thousand and twenty-six hundred years ago. During the wetter periods the forest grew and coalesced over wide areas, as it is at present. During the dry periods it shrank to small remnants surrounded by open, drier savannahs and scrubby forests. These forest refuges were concentrated along the base of the Andes, the Guiana highlands to the north, and smaller mountain masses to the south (fig. 1.6). They were determined primarily by topographic features that affected the movement of moisture-bearing winds, so the same refuges probably formed in each of the dry periods. The existence of these refuges is indicated by present-day regions of unusually high precipitation, and the distribution patterns of lizards, butterflies, and birds, including blackbirds.

Climates are, of course, still changing. South America can be expected to undergo other cycles in the future, but the fastest changes occuring now are those induced by the rapid expansion of the human population, which has at its disposal many means of modifying and changing the surface of the earth. The Amazon Basin forest is rapidly being cut and, except for western Brazil, may be reduced to small remnants early in the next century. The bare slopes of the Andes, long since deprived of their unknown native forest by centuries of human occupation, are being planted with pines and eucalyptus. Even in the lowlands, pines are replacing other trees because north temperate technologists know how to handle pines, but are baffled by the varied properties of the woods of tropical trees. Rather than

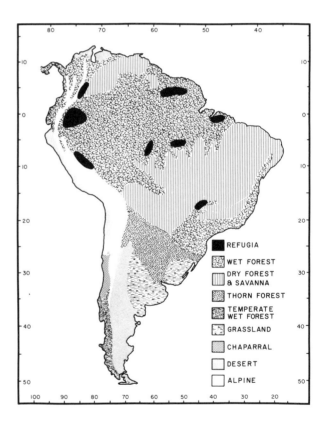

1.6 Probable location of forest refugia in South America during interglacial periods.

learn about tropical woods, we have chosen to replace them with something more familiar.

Over most of the tropics and subtropics the seasonal patterns of rainfall are determined primarily by the movement of the earth with respect to the sun. The higher the angle at which sunlight falls, the greater the amount of solar energy arriving at the surface per day. Solar energy heats the air near the ground and causes it to expand and rise. There is a belt of rising air located where the most intense heating occurs; that is, where the sun is directly overhead. But as air rises it expands, because it now occupies a greater volume and, expanding, it cools much as does air rapidly escaping from a balloon. Cool air can hold less moisture than warm air so, as air rises, clouds form and rain falls. Throughout the tropics and subtropics rainy seasons occur when the sun is directly overhead, while dry seasons occur when the sun is further to the north or south. At the equator the normal pattern is to have two rainy and two dry seasons per year. Farther away from the equator, in the subtropical zone, one of the dry seasons gets progressively longer and the rains coalesce into one season, during the summer portion of the year.

This simple pattern is complicated by the presence of mountains which, by forcing air masses to rise, also cool them and cause rain to fall. Thus the windward sides of tropical mountains are much wetter than the leeward sides, and rain may fall there during what would otherwise be dry seasons. Other modifications of the basic patterns are caused by proximity to or distance from the ocean and the temperature of the nearby ocean, resulting in a truly remarkable array of tropical climates—from extremely dry deserts to humid rain forests.

Since in the tropics and subtropics temperatures change so little, differences in amounts and timing of rainfall are the most important annual changes. Blackbirds, so many species of which are strongly dependent upon water for breeding and feeding, are especially sensitive to rainfall regimes and their effects on vegetation, rivers, lakes, and marshes. Some species, like the marsh-nesting ones, breed primarily during the rainy seasons, while some of the forest species breed during the relatively drier months when insects may actually be more abundant than they are during the rains. But, even at higher latitudes, rains may be important in determining breeding seasons.

Temperate Climates

In South America the Río Paraná flows southward, picking up water from the eastern lowlands of Bolivia and Paraguay and from the humid uplands of southern Brazil. It enters the Atlantic between Argentina and

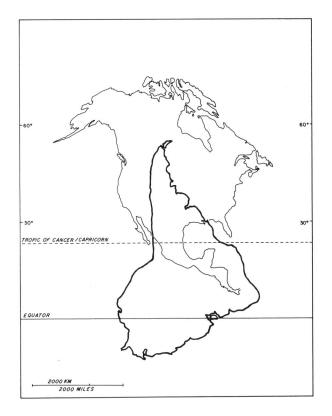

1.7 The fact that North America lies primarily in the temperate zone while South America lies primarily in the tropics can best be seen by rotating South America on the equator and shifting it to the west so that its temperate projection is centered over North America.

Uruguay as the broad Río la Plata, River of Silver, where it carries its muddy load far out to sea. The river has played an important role in the history of southern South America, as recorded in the rich folklore and folk music of the region. In the southern summer, fed by heavy rains in the subtropical portions of its drainage basin to the north, it swells and overflows its banks, creating vast marshes in its broad valley. The marsh birds of the valley mostly breed in the middle of the summer. A spring visit to the region will reveal little sign of breeding even though temperatures are quite favorable—in fact, more comfortable than they are in the very hot, humid summers.

In contrast, farther south where the spring winds of the Pampas are raw and the trees are still clinging to their winter buds, breeding activity is already under way. By late July (mid-winter) the southern screamers and black-necked swans are laying their eggs and ducks are courting actively. In August a number of ducks begin to lay and by September the small song birds of the marshes, blackbirds among them, can be seen carrying nesting materials. By January, when breeding is just reaching its peak further to the north along the Río Paraná, the southern birds have fledged their young and are molting their feathers in preparation for autumn.

The mirror image of these events occurs in the transition between subtropical and temperate North America. The red-winged blackbirds of Washington State have deposited their eggs before their Florida cousins, waiting for the late spring rains, have started to nest. Costa Rican redwings wait for the heavy rains of May and June to fill their breeding marshes, most of which become parched during the long winter dry season.

Geological History of North America

North America, too, underwent great changes during the Pliocene and Pleistocene but, unlike South America, most of which lies well within the tropics, North America is overwhelmingly temperate (fig. 1.7). The land mass increases with latitude in North America but decreases with latitude in South America. Unlike the situation south of the equator during the ice ages, great sheets of ice covered and recovered much of Canada and the northern United States, and temperatures were very cold for hundreds of miles south of the actual ice. Variations in the quantity of precipitation created large rivers and huge lakes, especially in the currently semi-arid intermountain West where small remnants of once huge glacial lakes persist in the valley floors. Plants and animals now living in most of the continent are recent arrivals, counting their residencies in but a few thousand or tens of thousands of years.

Events in the Pacific Northwest, where I have carried out most of my work, were particularly cataclysmic. Continental ice sheets, riding over high mountain masses to the north, joined by montane glaciers from the Cascade Range, blocked the channel of the Columbia River. Held back by this ice dam, glacial Lake Missoula formed, covering vast areas of northern Idaho and western Montana (fig. 1.8). Its former shoreline still can be clearly seen high on the slopes of the mountains surrounding the Flathead Valley in Montana. Eventually, the ice dam broke and the lake emptied in one of the mightiest floods ever to scour the surface of the earth. Flood waters raced across the extensive Miocene lava flows covering the Columbia Basin of eastern Washington, carving out extensive channels and leaving behind stark cliffs of columnar basalt. Today most of these channels are dry and are known to local residents as coulees. At the height of the flood they contained waterfalls larger than any in existence today. The scouring of the floors of these channels left many undrained depressions which, in the warmer and more arid times to follow, became the sites of vernal pools.

The Columbia River has returned to its former course and its basin, lying in the rain shadow of the Cascade Mountains, is arid to semi-arid. However, the rich soils and the availability of Columbia River water have made the region an attractive one for "reclamation," that is, its conversion from sagebrush desert and arid grassland to irrigated farmland. Water has returned in great quantities to the basin, but its flow is not as cataclysmic as it was during the great floods. In addition to creating permanent rivers where there were formerly only intermittent streams, irrigation water, applied at rates much greater than needed to sustain crops, has contributed to a raising of ground-water levels. This "perched" ground water is held there by the impervious layers of lava beneath it.

Rising ground-water levels created many new lakes with abundant supplies of insects, which attracted large numbers of blackbirds. These blackbirds, in turn, have attracted students of animal behavior and ecology who have taken advantage of the combination of desert climates and productive marshes to delve into the adaptive significance of blackbird social systems. My research career would have been very different if glacial Lake Missoula had not formed and there had been no great flood.

Modern Times

Climatic changes are still occurring but the most profound recent changes in the environments of the Americas have been due to human activity. We have cut forests, drained marshes, plowed grasslands, dammed

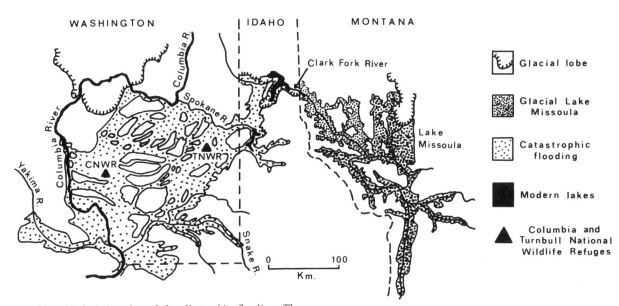

WASHINGTON | IDAHO | MONTANA

Columbia R.
Spokane R.
Clark Fork River
Columbia River
Yakima R.
Snake R.
CNWR
TNWR
Lake Missoula

0   100
Km.

Glacial lobe

Glacial Lake Missoula

Catastrophic flooding

Modern lakes

Columbia and Turnbull National Wildlife Refuges

1.8 Glacial Lake Missoula and the effects of its flooding. The marshes on the two refuges where I have carried out most of my research on blackbirds, the Columbia National Wildlife Refuge (the Potholes) and the Turnbull National Wildlife Refuge, were created by these floods.

rivers, and introduced many new chemicals into the environment. These changes profoundly affect where blackbirds can live today and where they are likely to be able to live tomorrow. Some species, particularly those that thrive in disturbed environments, have benefited greatly and are expanding their ranges as suitable habitats have been created in previously uninhabitable areas. Others, particularly those of mature forests, have found their habitats greatly reduced in extent. These processes can be expected to continue into the future as more and more people attempt to extract resources for living from a nonexpanding earth. This book, then, is a status report, both of knowledge about blackbirds and of their success in dealing with a species whose activities increasingly mold the features of the earth to its own ends.

2.1 *A scarlet-headed blackbird gapes into a bulrush stem while its mate watches.*

2. Gains from Gaping

Blackbirds, like most organisms, are influenced by many different events in their physical and biological environments that affect their chances of surviving and reproducing. The adaptive radiation of blackbirds into so many species and their ability to exploit a wide variety of habitats and prey types may be owing to a single factor—their unusual way of feeding. Their secret is finding prey not available to other birds. Yet this remarkable technique is so subtle that it is easily missed if one does not watch them carefully.

Not all suitable prey meet the eye. Many are hidden beneath objects or in crevices and holes. Nonetheless, most birds are superficial foragers. Some of them do extract prey from crevices with long bills. Woodpeckers dig into living and dead wood and probe with their tongues. In New Guinea and Australia, where there are no woodpeckers, large parrots are the chief diggers into wood. The shredded perches in the cages of Australian parrots in zoos signal their propensity to attack wood with their powerful hooked bills.

But blackbirds are more subtle than that. Along with members of two other families of birds—the starlings (Sturnidae), and crows, jays, and magpies (Corvidae)—they are able to open their bills forcibly against considerable resistance, a technique known as *gaping*. Gaping helps in finding prey in a remarkable variety of places.

Where Is Gaping Useful?

Meadowlarks insert their bills into the soil, force them open, and then extract prey. Burrowing spiders are especially vulnerable to capture by this technique. Meadowlarks are also adept at gaping into the bases of clumps of grasses, particularly bunchgrasses, which are favorite hiding places of many nocturnal insects during the day. Insects and other arthropods hide in the sheathing leaf bases of many aquatic plants where they can be exposed by gaping. Red-winged and yellow-headed blackbirds obtain much of their food during the spring and summer from these sites, and in South America the yellow-winged blackbirds do the same. Scarlet-headed blackbirds have powerful, chisel-shaped bills with flat ridges down the center of the upper mandible. They forage by hammering into the stems of such plants as bulrushes, forcefully opening the stems to reveal prey living within them (fig. 2.1). They must have some prior information about which stems to

split, because I randomly opened hundreds of bulrush stems in marshes in Argentina in areas where scarletheads were feeding but found very few insects. Obviously I failed to discover the subtle clues that aided the birds in their search.

Many nocturnal animals hide under objects during the day. Redwings, cowbirds, yellowheads, and Brewer's blackbirds all regularly seek prey by flipping over stones, dung, and small branches lying on the ground (fig. 2.2). I have also watched redwings foraging in riffles of small streams, turning over small stones by gaping in order to expose aquatic insect larvae attached to their undersurfaces.

Other plant structures provide hiding places from which prey can be captured only by an accomplished gaper. Dead twigs and branches often harbor large numbers of small insects, especially ants. They can be readily exposed by inserting a bill and forcing the branch open, especially if the branch can also be held firmly by the feet. This is the prime method of foraging of the solitary black caciques in tropical forests (fig. 2.3). In addition, they are fond of splitting slender stems of the bamboos that are so common in montane forests in Central and South America. The yellow-billed cacique has one of the broadest altitudinal ranges of any Central American bird. In Costa Rica it is common in second-growth tangles at sea level, and is also one of the most conspicuous birds in montane bamboo thickets as high as 3,500 meters. Everywhere it feeds by gaping.

Pine cones provide another opportunity for gapers. Yellow-backed orioles spend a great deal of time gaping beneath the scales of cones in Mexico and Central America. Gaping is also an effective means of foraging in clumps of pine needles. Many insects hide in the bases of these clumps during the day. Melodious blackbirds also gape into needle clumps in pine trees in Central America. Bark can be pried loose by gaping movements, and most of the arboreal icterids exploit that substrate at least occasionally. Orioles are the most frequent gapers into bark, but caciques and oropendolas are others.

Rolled and curling leaves, both living and dead, harbor many insects and are especially attractive to caciques and oropendolas. At La Selva, Costa Rica, I have often watched the scarlet-rumped cacique forage by grasping leaves with its bill and turning them over. If the cacique spots a prey item, it holds the leaf with its feet while extracting the prey with its bill, often by gaping. The smaller of the two Central American oropendolas, the chestnut-mantled oropendola, is also a great gaper into leaves. These birds sidle out to the ends of larger branches and gape into curled leaves or between two or more touching leaves. According to my observations, they prefer large-leaved species of trees, probably because these offer hiding places for larger prey.

Epiphytes growing on the trunks and branches of trees offer many excellent opportunities for gapers. New World forests are especially favorable because the dominant epiphytes are bromeliads (pineapple family), whose members have broad, succulent leaves with overlapping bases that hold water (fig. 2.4). These small aquatic communities high in the trees support animals as large as small frogs and some of the largest damselflies, which are known to breed only in bromeliads. Oropendolas, caciques, and orioles are the main probers of bromeliads on the mainland. The highly specialized Jamaican blackbird forages primarily in bromeliads in the cloud forests of Jamaica. Mosses are less suitable as places into which to gape, but I have seen the small golden-winged cacique gaping into the thick layer of mosses covering the trunks of alder trees in dripping cloud forests of northern Argentina.

Flowers, especially deep, tubular ones, offer resources of nectar and insects attracted to it that are available to birds with long, slender bills. Much of this resource can be extracted simply by probing into the flowers, as orioles often do, but they commonly open the flowers further by gaping. Such foraging by orioles can be highly destructive to the plant. I have watched Altamira orioles in the state of Tabasco, Mexico, destroying hundreds of blossoms of *Erythrina*, a legume commonly planted as partial shade over coffee and cacao, by pulling off the flowers and sucking nectar from the exposed basal nectaries. Their taste for sweets attracts orioles to feeders providing sugar water (artificial nectar), a reward normally offered to hummingbirds. Species recorded at hummingbird feeders in the United States and Mexico include Scott's, hooded, Baltimore, Bullock's, Wagler's, and streaked-backed orioles. Icterids seen feeding at flowers include Montezuma and chestnut-headed oropendolas; scarlet-rumped and yellow-billed caciques; the melodious blackbird; and orchard, hooded, Scott's, Baltimore, and yellow-backed orioles.

Virtually all the habitats exploited by icterids provide some opportunities for gaping. In grasslands there are objects lying on the surface, in holes in the ground, and in the bases of grass clumps. In marshes there are sheathing leaf bases, hollow stems, and floating mats of vegetation. In forests there are bromeliads and other epiphytes, curled and touching leaves, dead branches, and bamboo thickets. Pine forests provide cones, bases of needle clumps, and, in the tropics, bromeliads. In desert areas there are flowers, especially the large

2.2 A male yellow-headed blackbird uses a combination of gaping and an upward movement of his head to flip over a piece of cow dung.

2.3 *In the understory of a wet tropical forest, a solitary black cacique searches for food by gaping into a dead twig.*

flowers of yuccas, and the leaf bases of succulents. In streams there are small stones and floating mats of vegetation. Most species of icterids that have been studied more than casually have been observed to use gaping movements while foraging, though the importance varies greatly from species to species.

It is probably not an exaggeration to say that development of the ability to gape may have been the key change responsible for the remarkable adaptive radiation of the blackbirds. Virtually all forest species seem to rely on gaping to obtain much of their food, and it is this method of foraging that distinguishes them from other birds foraging in the canopies of trees and shrubs. Gaping makes accessible to them food not available to species unable to gape, and the places in which they gape house larger prey than is usually available on the surfaces of those same structures.

How Feet Help Gaping

The value of gaping is increased if a bird can coordinate movements of its bill and feet, just as it takes one hand to hold a banana and another to peel it. Many blackbirds hold a branch, leaf, flower, or other object in one

or both of their feet while gaping it open with their bills. This enables them to reach out for leaves on branches that are too slender to support their weight, pull the branch back, and hold it in their feet on a more substantial one. Even so the arboreal blackbirds foraging at the tips of slender branches often find themselves on branches too slender to support them in an upright position. They often hang almost upside down as they hold branches and gape into leaves. While this is not a particularly elegant way of getting breakfast, judgment should be based on success rather than appearance. It works well and so it is used.

The ability to coordinate movements of bill and feet, and to hang upside down at the tips of branches, makes the arboreal icterids especially well suited for forest edges and savannas where, because the tree canopies usually do not touch, there is a broad vertical range of leaf-bearing branches in which the birds can forage. Most of the orioles are not birds of closed forests. Rather they occur in savannas and disturbed forests and along rivers where an abundance of foraging places are available. Holding food in the feet has also been reported in some primarily terrestrial species such as the melodious blackbird, Brewer's

2.4 *A chestnut-mantled oropendola forages in the base of the leaves of a bromeliad.*

blackbird, several species of grackles, and the brown-headed cowbird.

The ability to forage in this manner may have had still another influence on blackbird evolution. Many of the arboreal species have displays in which the bird swings forward until it is actually upside down, head pointed down and tail up on the branch. In the more vigorous displays the birds lose their balance and must flap their wings to regain a normal upright position. Perhaps the tendency to do this was enhanced by the already well-developed ability to forage in similar upside-down positions.

Adaptations of Blackbirds for Gaping

Gaping, then, gives blackbirds a distinct foraging advantage, but what are the anatomical features that allow them to perform this unusual act successfully? Most birds have powerful muscles for closing the bill so that they can grasp and crush prey. Normally, however, bills are not opened against strong resistance, and gravity can be counted on to do most of the work. Muscles for opening the bill are small in most birds. Gapers, on the other hand, open their bills against the resistance of soil, wood, rocks, and other objects and generate considerable force to expose prey in these situations. As a result, gapers have unusual head muscles in comparison with other birds.

From an engineering standpoint, the avian skull is a rather remarkable structure. The upper mandible of a bird, unlike that of most vertebrates other than reptiles, can be moved up and down on its articulation with the cranial part of the skull. This is most readily demonstrated by parrots when they are feeding, because their skulls permit an unusual amount of movement between mandible and cranium, but all birds have this ability to some degree. This movement of the mandible on its hinge is communicated by a series of rodlike bones to a bone known as the quadrate, which occupies a key position in the skull (fig. 2.5). Movement of the quadrate, primarily as a result of the contraction of a muscle running between the quadrate and the inside surface of the jaw, pulls the upper mandible downward. Upward movement (protraction) of the upper mandible is accomplished by an opposing set of muscles.

As can be seen from Figure 2.5, the apparatus is rather elaborate. One function of the arrangement is to provide coordination between the movements of the upper and lower mandibles. The result is that the upper mandible is protracted only when the lower mandible is depressed, and lowered whenever the lower mandible is raised. This not only increases the angle of the gape when the bill is open; it also preserves the axis of the bill so that it is always in the same position relative to

the rest of the skull. If the axis were to change, it would be very difficult for birds to snatch moving prey quickly and accurately.

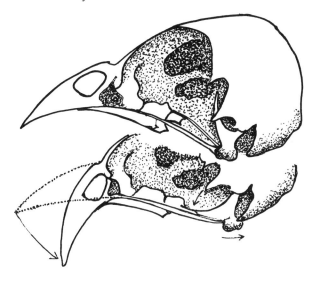

2.5 *The position of the quadrate when the upper mandible is protracted (above) and when it is retracted (below).*

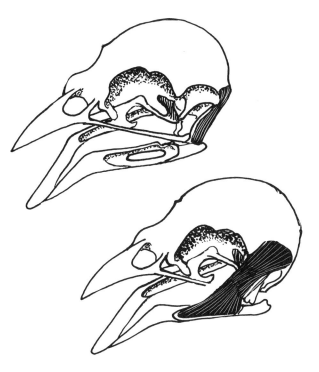

2.6 *The lower mandible of birds is lowered by muscles running from the rear end of the mandible to the back of the skull (upper) and raised by muscles attached to the main part of the mandible and to the side of the skull (lower). From the locations of the muscles it is easy to see why their contraction and, hence, shortening, moves the mandible as it does.*

Many muscles are involved in movements of the skull and, unfortunately, they have very complicated names. Luckily, however, the essentials can be grasped if we deal simply with sets of muscles grouped according to their general functions (fig. 2.6). There is a single muscle on each side for depressing the lower mandible, and a single muscle for elevating (protracting) the upper mandible. In addition, there are three retractors that simultaneously pull the tip of the upper mandible downward and draw the lower mandible upward. Finally, the adductors of the mandible have the sole function of elevating the lower mandible and bringing it into contact with the depressed upper mandible for seizing and crushing prey.

The most conspicuous adaptation in the musculature of blackbirds is the great development of the muscles for protraction. The muscles for elevating the upper mandible and those for lowering the lower mandible, which are small and weak in most birds, are very well developed in the strong gapers (fig. 2.7). In powerful gapers such as the meadowlarks, these muscles are larger than the muscles that close the mandibles, whereas for the common grackle, which crushes hard nuts, such as acorns, adductor muscles are more powerful than depressors.

For a bird exposing hidden prey by gaping movements of its bill, it would be highly advantageous to be able to see directly into the space created by the open bill. The skulls of gapers and the positions of their eyes allow them to do precisely that (fig. 2.8). The shape of the bill and its projections onto the skull are such that a direct line of sight is preserved from the eyes to the tip of the bill. The axes of sight of both eyes run parallel to the lower edge of the upper mandible, which gives the blackbirds their unusual and characteristic appearance when viewed straight on.

Visibility down the axis of the bill is further enhanced by the presence of special feathers between the eye and bill (fig. 2.9). They are velvety and black in most icterids, even if the head is generally brighter (fig. 2.10). The black bristles probably serve to eliminate the disturbance that would be caused by light reflecting from those feathers—not unlike the football receiver who applies charcoal to his eye sockets to reduce glare. In some blackbirds, such as the oriole blackbird shown in Figure 2.10, this zone is actually featherless, as is also the case with some other birds, such as falcons, that need acute vision directly in front of them.

Blackbirds Are Compulsive Gapers

The history of blackbirds is to a large extent a story of what the powerful opening of their mouths has done for and to them. Many blackbirds have such a predilection for gaping that they seem unable to stop

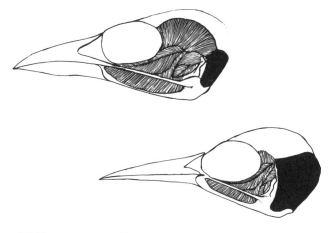

2.7 The common grackle (above), which crushes hard nuts and seeds but does relatively little gaping, has powerful crushing muscles but weak gaping muscles. In contrast, meadowlarks (below) obtain much of their food by gaping and, consequently, their muscles for opening the bill (shown in black) are larger than those for closing it.

2.8 A brown and yellow marshbird, viewed from directly in front, its eyes positioned so that it can see into the space at the tip of its opened bill.

doing it when it serves them no end. Redwings gape vigorously in piles of seeds as if to expose hidden grains. I once had a captive redwing that went around its cage gaping into any crevice or space between objects it could find. One of its favorite pastimes was to walk on my hands and insert its bill between my fingers and gape them apart. I was amazed at the force such a small bird could generate, just as I am surprised at the force blackbirds can generate when they bite. A meadowlark raised by the famous ornithologist Margaret Morse Nice went around its cage gaping into every space it could find, enlarging all existing holes in papers and fabrics. Captive starlings and crows behave in the same way. These birds reveal what has been so important in their evolution by performing their act whenever they have an opportunity.

2.9 Black and bristly feathers enhance visibility down the axis of the bill. Shown here, from top to bottom, a western meadowlark; a melodious blackbird; a starling, a member of a different family of birds but a species that obtains much of its food by gaping; and a merlin, a falcon that pursues and captures flying birds.

2.10 The area between the eye and bill of most blackbirds is black, regardless of the color of the rest of the head. Shown here are the oriole blackbird (whose black area is actually bare skin), the streak-backed oriole, the yellow-backed oriole, and a female red-winged blackbird.

3.1 *This female yellow-headed blackbird has just captured a large dragonfly which she will deliver to her nestlings.*

26

3. Finding and Capturing Food

Not unlike us, blackbirds are preoccupied with food much of their lives. The need for food and the problems associated with getting it vary with age and environmental conditions. Young blackbirds do not seek their own food because they are fed by adults until they are almost fully grown (fig 3.1). Among the smaller species this period may be no more than a few weeks, but it is critical to their lives. Starvation is an important source of mortality among nestlings in some intensively studied species.

For adult blackbirds the breeding season is a time of special food needs and stresses. There is the extra demand of establishing and defending territories, which may occur early in the spring when temperatures are still cold and food supplies are low. Females eat extra food to allow them to form eggs and build nests. During incubation, when the eggs must be covered most of the time, there are pressures to accomplish feeding in as short a time as possible. Especially great demands come when the young hatch because, during their periods of rapid growth, a single adult may have to gather up to six times as much energy per day as it would need if it were only feeding itself.

In most birds the breeding season is followed by a molting period, when the old feathers are replaced. This occurs over a period of several weeks, and also imposes an energy strain. At no time are the birds not well covered with feathers, but their flight is often impaired and they may look very scruffy.

Migration, for those species that perform it, is another energy-demanding time. It is preceded by a period of overeating during which the birds deposit excess energy as fat in special areas underneath their skin. Fat deposition is an interesting phenomenon because normally, even when presented with unlimited food, birds regulate their intake very precisely to maintain a relatively constant body weight. They seem to know, without benefit of a Surgeon General, that excess weight is costly. For birds, in addition to the health liabilities of carrying fat deposits, there is probably a predation risk associated with obesity. Birds with lots of fat may be less maneuverable and, hence, easier for predators to catch. Just prior to migration, however, birds put on enormous amounts of weight to fuel their long flights.

Not only do needs for food change, but the availability of food changes diurnally, seasonally, and between years. Although food is not in short supply most of the time for birds, the population sizes of many birds are determined by the availability of food. Blackbirds are probably no exception, but as yet no species has been studied in sufficient detail for us to say with confidence what regulates its population size. The experience of many dedicated field ornithologists over several decades shows that determining what regulates a population is one of the most difficult tasks any ecologist can set out to accomplish.

Even if we do not know precisely what controls the numbers of blackbirds, striking changes in the abundance of some species in recent decades can be correlated with changes in food supply. Most of these changes are caused directly or indirectly by human activities.

The red-winged blackbird may be the most abundant bird in North America today, but it has not always been so. Part of the increase in its population may be owing to greater productivity of lakes and marshes, because of fertilizer inputs from surrounding farmlands. More important, however, is the conversion of vast upland tracts from natural grasslands and forests into croplands. These croplands are difficult for most birds to exploit but, as most farmers know, they often harbor large populations of insects. In recent decades redwings have adapted to upland environments and are now breeding extensively in agricultural fields in the East, Midwest, and parts of the Far West, especially California. Population densities are not as high in upland habitats as they are in marshes, probably because of lower food densities; but because of the large extent of croplands compared with marshes, the proportion of the population of redwings that now breeds in uplands is quite high.

In addition, human activities have greatly increased the food available to blackbirds in winter. Modern harvesting practices leave substantial residues of seeds in fields, some of them from the crop plant itself, others from weeds. These fields are ideal foraging places for blackbirds and have greatly increased overwinter survival. For example, the winter range of the yellow-headed blackbird has greatly expanded as a result of this newly available winter food supply. Until recently, yellowheads were restricted as wintering birds to the Great Valley of California, south and east to southern Arizona and the southern end of the Mexican Plateau. Today, however, there are thousands wintering in the Columbia Basin desert of central Washington, hundreds of miles north of the previous limits to their winter

3.2 *The shiny cowbird.*

range. They are sustained by crop residues and grain at cattle-feed lots.

The most spectacular recent changes in range and abundance have occurred among the cowbirds. When Europeans first settled North America, the brown-headed cowbird occurred primarily in the Midwest and Great Plains, particularly in association with herds of bison. The insects flushed by grazing mammals are an important food for the birds. In the West the brown-headed cowbird was widely distributed but relatively rare. Europeans both cleared the countryside and introduced domestic livestock in large numbers, thereby changing foraging conditions for cowbirds over much of the continent. As a result, cowbirds have become far more abundant than they were formerly and have greatly extended their breeding range. They are now common breeding birds throughout the deciduous forest zone of eastern North America and have become common in most of the region west of the Rocky Mountains.

As late as 1940 the naturalist Ira Gabrielson reported the cowbird to be uncommon in most of Oregon and only a rare straggler west of the Cascades. In fact, none of the early naturalists found cowbirds anywhere in Oregon. The first report came in 1902. Cowbirds are now abundant throughout the Pacific Northwest, including the humid coastal strip from which they were virtually absent less than thirty years ago.

A similar spread has been occurring in the South American shiny cowbird (fig. 3.2) which has been extending its range southward through the temperate parts of South America. It now occurs as far south as Patagonia in Argentina and is spreading rapidly in Chile. We are not sure whether this bird crossed into Chile from Argentina on its own or was accidently introduced. The latter is possible because the males of this species are excellent songsters and are often kept as caged birds in Argentina. In any case, the first reports of shiny cowbirds in Chile around 1905 came from the Central Provinces where the mountain barrier with Argentina is very high and formidable to a basically lowland bird. Today the shiny cowbird is quite common in Chile and ranges from the Copiapó Valley in the Atacama Desert in the north, south to Ayasen in the humid beech forest zone of southern Chile. As elsewhere, it frequents cultivated areas and pastures, and commonly feeds in close proximity to domestic livestock.

Even more spectacular, however, has been the shiny cowbird's northwest movement through the West Indies. The first such specimen was collected on Vieques Island near Puerto Rico in 1860, but it was not recorded again anywhere in the islands until 1899. Since then it has been spreading northward and has now reached Hispaniola (fig. 3.3). The rate of progress

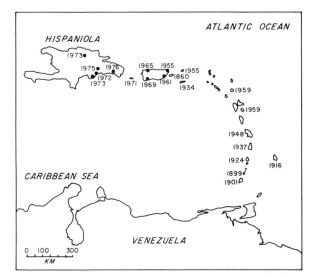

3.3 The spread of the shiny cowbird northward through the West Indies.

northward through the Lesser Antilles has been slow, on average about 8.3 kilometers per year. It is still not present between Antigua and St. Croix, a distance of 300 kilometers. In the Greater Antilles, the shiny cowbird has spread rapidly westward from a population center in the Virgin Islands and eastern Puerto Rico, probably the result of a more recent invasion, since no birds were seen in that area between the 1860 observation and 1934. The range expansion seems to be continuing; cowbirds may well reach Cuba within a few years. There is no reason to expect the northward spread to stop there. We may actually have an opportunity to observe what will happen when the shiny cowbird comes into contact with the brown-headed cowbird and both species lay their eggs in the nests of the same hosts (see Chapter 7).

The Foods of Blackbirds

So many different types of food have been recorded from diets of blackbirds that it is easier to list what blackbirds don't eat. Virtually all kinds of terrestrial and freshwater arthropods are eaten by one or more blackbirds. Snails are prominent in the diets of some marsh-nesting species. Blackbirds eat the seeds and fruits of hundreds of species of plants and the more robust birds are able to handle large seeds. Grackles can crush acorns, with the prominent keel in their powerful bills.

Even other vertebrates find their way into the stomachs of blackbirds. Grackles are adept at catching emerald shiners (*Notropis atherinoides*) in shallow water. William J. Hamilton found remains of fish in 70.8 percent of stomachs of nestling grackles in New York. Fish constituted 6.6 percent of the bulk of the food present in those nestlings. In addition, 10.8 percent of them contained amphibians. The scarlet-headed blackbird occasionally brings frogs to its nestlings in Argentina, and several species eat lizards. I saw a melodious blackbird with a small lizard in its bill in Peru in 1974. Little is known about the prey of large oropendolas in the canopies of tropical forests, but they probably find many frogs and small lizards in their probings in and around the large bromeliads that festoon the branches.

Blackbirds have even been observed to attack other birds. Carl Helms watched a redwing attack, kill, and sever the head of a sharp-tailed sparrow that was caught in a mist net. During a cold snap in the spring of 1972, rusty blackbirds were observed to attack other birds several times near Fort Good Hope in the Northwest Territories of Canada. The attacks occurred after a severe blizzard on May 21-22 when food may have been in short supply. The birds attacked included tree sparrows, white-crowned sparrows, and Lapland longspurs, but only longspurs were actually seen to be killed and eaten. Common grackles have been seen capturing and eating small birds. Despite these cases, other birds and mammals are unusual items in blackbird diets. The blackbird world runs primarily on energy from arthropods, seeds, fruit, and nectar.

Diets through Time

Food availibility changes seasonally and so do food needs. Seeds are excellent foods for adult birds but are poor nestling food because they have little water and are difficult to digest. Insects may be unavailable in cold weather. Fruits occur in their appropriate seasons. The diets of most blackbirds shift in response to these changing needs and availabilities.

Among temperate-zone blackbirds, the most common pattern is a shift from reliance upon seeds during the fall and winter to a reliance on arthropods, chiefly insects and spiders, during spring and summer. The shift is not complete, but the changes are striking. For example, the bobolink breeds in wet meadows and pastures in North America from the Rocky Mountains east to the Atlantic Coast, with isolated small populations farther west. During the breeding season the young are fed almost exclusively on arthropods such as spiders, moth larvae, and grasshoppers, which also dominate the diets of the adults at that time. After breeding, the birds migrate southeastward to the southern Atlantic Coastal Plain where they feed extensively in cultivated rice fields and earn the name of ricebird from the local residents, many of whom do not recognize it as the same bird that passes through the area in spring with a different plumage. In April, the males in their brilliant black and white plumages

are strikingly different from the drab brown ricebirds of October. After fattening themselves on rice the birds fly to Argentina where they spend the winter (actually the southern summer) in the pampas, eating a mixed diet of seeds and arthropods.

Early in spring, redwings subsist primarily on seeds, many of them exposed by melting snows. The marshes in which they nest may already have thawed, but it will be some time before the emerging aquatic insects upon which they feed their young first appear. At this time, because of the lack of food in their nesting areas, the birds occupy their territories only briefly, defending them for short periods at dawn and dusk, but spend most of their days foraging for seeds in adjacent uplands (fig. 3.4).

As spring progresses, however, food supplies slowly increase on the marshes and the need to be there to defend future breeding space increases. As a result the time spent on territories gradually increases until by the time females arrive and are selecting their nesting places the males spend nearly all day on their territories. Gradually, also, the proportion of arthropods, principally insects, increases in their diets until by the time nesting is underway both males and females subsist almost exclusively on animal prey.

Animals continue to dominate the diet of redwings through the time that molting takes place in late summer. At that time the birds fly with some difficulty because of missing feathers, so they tend to remain close to the thick, coarse cover of the marsh grasses and cattails in which they can easily hide if pursued. Fortunately, insects are still abundant in the marshes at that time and finding food is easy.

Once molting has been completed, redwings gather into large flocks and range much more widely. At this time they become pests in crops, especially corn and rice. Following harvest, there is a gradual shift south, with most of the population wintering south of the areas of regular snow cover. During the winter they feed primarily upon crop residues and weed seeds until the increasing daylight and rising temperatures of early spring once more trigger the annual north migration to the productive marshes and croplands of higher latitudes.

Although it is always warm in the tropics, the striking seasonal changes in rainfall have profound effects on the abundance of different types of food. In areas with pronounced wet and dry seasons many forest trees lose their leaves during the dry season. Even in regions where, because of the shortness of the dry season, most trees are evergreen, there is a peak in production of fresh young leaves at the beginning of the wet season. Because young leaves are a better food source for most leaf-eating insects than are older, tougher leaves, the annual cycles of many insects are

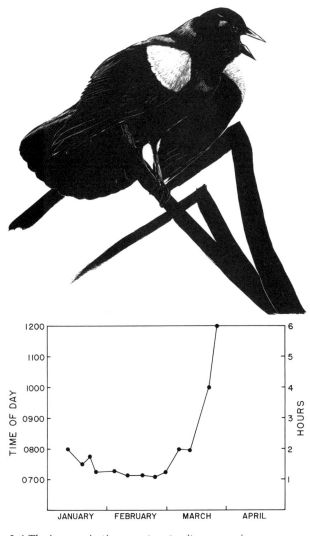

3.4 *The increase in time spent on territory as spring progresses for male redwings in a marsh at Berkeley, California.*

adjusted to these peaks in availability of young leaves. Over much of Central America the long dry season begins in November or December and lasts until late April or early May. Most of the resident birds, among them orioles, begin to nest toward the end of the dry season and are feeding their young during the period of maximal insect abundance early in the wet season.

Flowers, like leaves, are present throughout the year in the tropics, but they peak in abundance in the dry season in most localities. Dry-season flowering is advantageous to plants because the best time to disperse fruits is at the beginning of the rainy season, when the soil is moistened and good germination sites for seeds are available in places disturbed during the long dry season. Few flowers and fruits are present in the latter part of the rainy season in most tropical areas.

Most tropical plants surround their mature seeds with a nutritious, fleshy fruit. Therefore, most eaters of

seeds are also eaters of fruits. Human activities, however, increase the abundance of hard-coated seeds that drop directly from the plants to the ground, because fruit-laden forests and woodlands are replaced by artificial grasslands whose dominant plants do not produce fleshy fruits. Blackbirds are conspicuous in these grasslands in many parts of the neotropics, feeding on the abundant seeds found on the ground in those areas.

In addition to being sensitive to seasonal changes in food availability, blackbirds opportunistically exploit unusual abundances of food whenever they occur. During those springs when cicadas are abundant in the sagebrush deserts of eastern Washington, female redwings locate the singing male cicadas by their loud, chirring sounds. Cicada is a regular treat for nestlings in those years.

Blackbirds are among the main exploiters of grasshopper outbreaks in the American West, joining birds of many different kinds to fatten themselves on those energy-rich morsels so readily available in some late summers. These outbreaks usually come too late to contribute to nestling food supplies, but they may improve overwinter survival of the birds, which are able to enter the winter with better fat reserves than in years of no grasshopper outbreaks.

Agricultural activities can create mini-outbreaks of insects. In California, for example, irrigated pastures are regularly flooded during the growing season. Each time that happens the insects that normally hide during the day among the bases of the grass clumps are forced to climb to the upper portions of the leaves and stems—the only parts of the grasses remaining above water. At those times, blackbirds, especially the highly colonial tricolored blackbird, congregate in large numbers in those pastures to harvest the unusually vulnerable prey items.

Harvesting creates another food bonanza by flattening the vegetation in which insects normally hide. In many parts of North and South America blackbirds flock to freshly cut croplands, often following close behind the harvesters, just as they follow plows in the spring.

How and Where Blackbirds Choose Their Prey

All birds face the same set of foraging decisions. They must decide where to forage, when to leave the place they are foraging to seek another site, which prey items from among those they encounter to capture, and how much to eat per day. To understand how these decisions are made ecologists have developed theories and have tested them by observing actual choices made by birds in the field or in the laboratory.

Developing such theories is a tricky business because one must first guess possible decision-making rules on the basis of reasonable arguments. What are reasonable arguments to us may not be reasonable arguments to the birds themselves. We may think energy is the important feature, but the birds may be looking for vitamins. Then the rules must be framed in ways that allow them to be tested. This is also difficult because we cannot ask birds directly why they did what they did. Their decision-making rules must be deduced from their choices. We must find situations where the patterns of their choices allow us to distinguish among various rules. The whole operation has some of the aspects of a detective game, which is, in part, why playing it is so exciting.

Choices are made on the basis of both the energy and the nutrient content of prey items. Like us, birds need vitamins and minerals, though exactly what they need may differ at different times of year. Also, toxic prey are normally avoided even if they contain lots of energy and nutrients. Unless one is extremely hungry it is not worth getting ill over a few calories. Monarch butterflies are toxic to birds because they contain poisonous cardenolides, and are generally avoided. Both the larvae and the adults of this large colorful butterfly obtain from their food, principally milkweeds, the cardenolides that make them toxic. Birds do not inherently know that monarchs are toxic. They try them first, become ill, and avoid them thereafter. Perhaps because of this toxic protection monarchs are able to migrate and gather in large wintering roosts in California and Mexico where temperatures are cool but not cold. Nonetheless, Linda Fink and Lincoln Brower of Amherst College found that birds accounted for over 60 percent of the mortality of overwintering monarchs at several sites in central Mexico. The two birds responsible were black-backed orioles (fig. 3.5) and black-headed grosbeaks.

How can these birds get away with eating large numbers of toxic butterflies? The key to the problem is that there is a great deal of variability in the level of cardenolides in the butterflies. The majority of the butterflies are only weakly toxic. Fink and Brower observed that the orioles released unharmed seventy-five of the ninety-eight butterflies they captured. Black-backed orioles also ate less of individual butterflies that contained more cardenolides than they did of less poisonous ones. The orioles could not tell more toxic individuals from less toxic individuals and so captured them at random. They then tasted each butterfly and released it if it was above a certain level of toxicity but ate some of it if the cardenolide concentration was below that threshold. How much of it they ate depended on how toxic it was. This is a smart way to behave because the orioles are very sensitive to the toxic substances found in monarchs. Eleven of them were fed powder from toxic monarchs and all of them vomited.

It may seem strange that the birds should bother to sample toxic prey and so risk a mild poisoning. The explanation is that the butterflies are present in great concentrations and are very easy to catch. Even if an oriole must catch a butterfly to tell if it is good to eat, the time required to find out is very short. Since many of the butterflies contain less than one emetic dose of poison, the birds can ingest quite a bit of energy safely. Morever, in mid-winter, a time of low abundances of other insects, monarchs are especially attractive.

But most insects are nontoxic, and insectivorous birds, particularly in the temperate zone, are more concerned with getting energy than with avoiding toxins. Therefore I started my detective's search for the blackbirds' decision rules by assuming that in general a blackbird would benefit by gaining as much food energy as possible while foraging. If it did so it would thereby reduce the time during which it was exposed to higher risks, mostly from predators, associated with foraging. It would also have more time to spend on other activities such as body maintenance, courting, defending a territory, or resting. If there are young to be fed, maximizing the rate of capture of food energy also permits the highest rates of delivery of food to the young and, hence, the largest number of young to be raised.

To maximize the rate of energy intake a bird must select the best foraging sites and those prey which, if taken, yield more energy than those not taken. It is not too difficult to specify the decision-making rules that lead to these results. To determine which sites it should visit, a blackbird (or any other animal) needs to know two things. First, it needs to know, or have an estimate of, how well it is doing in the place where it is foraging. Second, it needs to know how well it can do elsewhere. Clearly, no individual will have perfect knowledge about these possibilities, but an experienced individual may have enough informaiton at its disposal to compare alternative choices.

The way to evaluate decision-making rules can be illustrated by a simple graph that plots the amount of energy captured while foraging against time (fig. 3.6). For each site in which the individual might forage, its success can be plotted as a curve that shows the amount of prey energy the individual has captured as it forages. If that curve never rises as high as the line representing the overall average energy intake rate for the environment, the forager would be better off not visiting that patch. If the curve rises above the line, the patch should be visited because, for a while at least, the forager can do better there than it can, on average, elsewhere.

3.5 *A male black-backed oriole has captured a wintering monarch butterfly in the pine woods in the mountains above Mexico City.*

This tells us which patches should be visited, but when should the bird leave a patch to go elsewhere? Basically, the same rule applies. A forager should leave a patch when it can no longer do as well there as it could elsewhere. In the graph, this point is reached when the slope of the line showing foraging success in the patch becomes parallel to the line giving the overall average for the environment. Until that time the forager is doing better than average. After that point, it is doing worse. Like a smart blueberry picker, the bird should leave before it has exhausted the food in the patch.

Thus, the decisions about which patches to visit and when to leave them are given by a single set of rather simple rules, providing the patches are equally risky places in which to forage. The problem is not to determine suitable rules but to test whether birds actually use those rules and to determine how they estimate the curves they are required to know if they are to follow those rules. Birds do know a great deal about their local environments and how to best forage in them, but, before looking at the evidence, we need a bit more theory to guide us.

The slope of the straight line in Figure 3.6 assumes that the foraging individual is making the best choices from among the prey it is encountering. But what are the best choices? If an energy-maximizing decision-making rule is being followed, the value of a prey item is the ratio of the amount of useful energy (E) the predator can expect to get if it pursues that prey, to the amount of time (h) it takes to pursue and consume it. This ratio, which gives the rate of energy intake from pursuits of prey of that type, is not equal to the values that obtain if the pursuit is successful. What the forager actually gets if it pursues that type of prey depends on the proportion of pursuits that are successful. For example, if a prey item (grasshopper) contains one hundred calories but a blackbird captures it only half of the times it tries, then the average energy intake per pursuit is only fifty calories. Similarly, the average time of a pursuit depends on the fraction of pursuits that are successful and the lengths of successful versus unsuccessful pursuits. The actual calculation of the energy to handling time ratio (E/h) is very difficult, but we need not be worried about these details. We need only to know that we can rank all prey types by their E/h ratios from the best prey (highest E/h) to the worst prey (lowest E/h).

With this formulation we can pose the dilemma of the forager. Suppose it has encountered a prey item of some intermediate E/h. Should it eat it or reject it and continue to search for better prey? The answer depends upon how abundant the better prey are. If they are sufficiently common, the forager will find better prey often enough that its rate of energy intake will be higher if it rejects the prey in front of it and continues

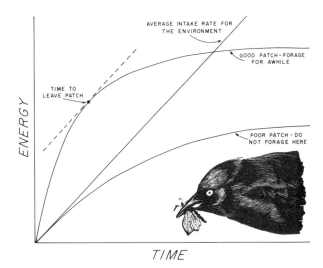

3.6 *The amount of energy a forager can obtain from two different patches, a good one (upper curve) and a poor one (lower curve). The average rate at which the individual can expect to find food in the overall environment is shown as a straight line from the origin of the graph whose slope gives the amount of energy harvested per unit time spent foraging. If the harvesting curve for a patch rises above this line, the forager should search for food there. If it does not rise above the line, the forager will do better to pass quickly through that patch to find a better one.*

to search. If better prey are not very common, however, the predator gains more energy if it takes the intermediate prey. Calculating how common the better prey must be for the predator to gain by rejecting an intermediate prey is complex, but there is a simple, useful rule of thumb. Whether or not a prey type should be eaten by a predator does not depend on its own abundance but entirely on the abundance of better (higher E/h) prey items. In other words, if good prey are common, the predator should reject all other prey. As good prey become less abundant, the predator adds less desirable prey to its diet while still taking the good prey whenever they are found.

The best tests of choice of prey by foraging birds have been made with titmice, but enough is known about foraging by blackbirds in the wild to tell us a great deal about their choice of places to forage and prey to take, even though we cannot provide rigorous tests of the theories themselves. The theories serve as guides to help us decide what to look for and how to interpret the results.

Blackbirds do reject toxic prey. In the deserts of central Washington where I have done most of my work, several species of common large black beetles are active during the blackbird breeding season. They have very unpleasant odors and are probably not good to eat. Even though they are common in places where blackbirds seek food, I never found one of them in over a thousand hours of sampling of foods delivered to nestlings. Similarly, moth larvae covered with long, irritating hairs are sometimes common on the cattails in the vicinity of redwing nests. These also never appear in nestling diets despite the fact that foraging birds must encounter them regularly.

Rejection of prey based on their size is a bit more difficult to document. The prey available to blackbirds in my study area ranges from large cicadas and dragonflies to tiny flies. All are regularly eaten. There appear to be no toxic species among the large group of aquatic insects that form the major parts of the diets of breeding blackbirds. There is a time during the day, namely the main emergence period of dragonflies and damselflies in middle to late morning, when large prey become especially abundant (fig. 3.7). If blackbirds were ever to reject small flies, the emergence period is when they should do so. My results are, however, quite clear on that point. Small flies are eaten throughout the day and are taken along with the larger prey even when capture rates of large prey are high. At first I was very puzzled, but the blackbirds were smarter than I was. When I calculated the capture rate of large prey that would be necessary to cause the blackbirds to reject the smaller prey, it turned out to be as high as or higher than the normal capture rates. This is because handling times for the recently emerged aquatic insects are so very short. Probably no more than a single second of foraging time is lost if the bird decides to pick up any of the insects.

The situation was different for redwings in a marsh in Costa Rica where they fed primarily on grasshoppers and spiders. These prey were more active and harder to catch. The birds could not bring more than one at a time to their young in contrast to the twenty or more insects that constitute an average prey load for Washington birds. I measured the sizes of grasshoppers the birds delivered to their young and compared them with the sizes of those that were on the marsh vegetation, as sampled by beating the vegetation with an insect net. The marsh was full of small grasshoppers but the birds brought almost entirely large ones to their young (fig. 3.8).

But what about selection of places in which to forage? To examine this problem we need to know not only how good the places are but also where they are. This is because birds are often constrained as to where they can go. During the nonbreeding season there may be fixed roosting sites to which birds return every night and from which they spread out during the day. During the breeding season adults may be flying to and from a nest containing eggs or nestlings. In these cases the distance from the roost or nest to the foraging site is clearly of great importance. Of two equally good foraging sites, the one closer to the nest will allow

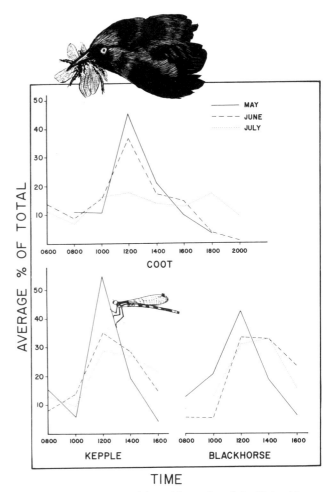

3.7 Emergence patterns of damselflies at Coot Lake (Columbia National Wildlife Refuge) and Kepple and Blackhorse lakes (Turnbull National Wildlife Refuge) in Washington State.

individuals to deliver more energy to their young per hour.

Gathering food that is not eaten on the spot but is delivered to a fixed site, either for storage or for offspring, is called central place foraging. The unit of interest is the round trip consisting of a journey out, a searching period, and a journey back. Energy is taken only during the search period but energy is expended in all three. In fact, expenditure rates are usually higher during the trip out and back than during the search because of the high costs of flight. A central place forager faces two important problems: where to search, and how large a load to gather before returning. Its choice of individual prey should be the same as in any other foraging case.

In some cases load size is easy to determine. The first prey captured so encumbers the predator that it cannot efficiently capture another and must return with the first one. In other cases, the first prey can be held in such a way that the predator can capture additional prey with equal ease. In such cases the predator should

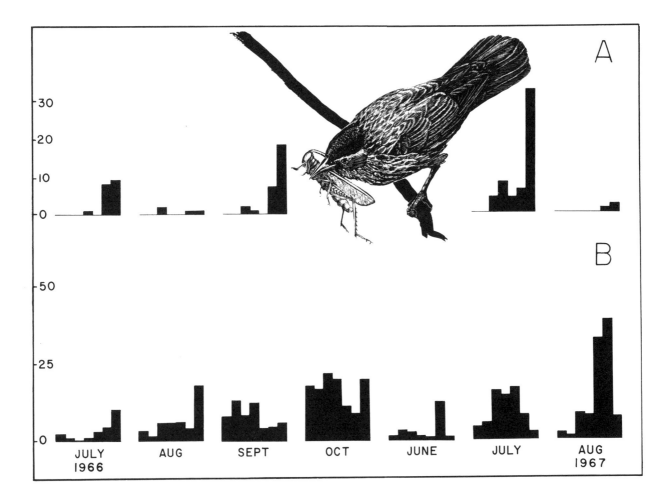

3.8 Numbers of grasshoppers of different sizes in a marsh in Costa Rica, plotted as frequency histograms, with the smaller sizes to the left and the larger ones to the right. The sizes as sampled by an insect net are shown below (B). The sizes of prey brought to the nests by the female blackbirds are shown above (A).

continue to load, but how long? Again, a simple graph helps us visualize the problem (fig. 3.9). As before, we graph energy against time, but now we also need to account for the traveling time during which no food is captured. To do so we start the line representing the rate of energy delivery not from the origin of the graph but to the left by a distance equal to the round trip traveling time from the central place to the foraging area. The curve showing accumulated energy rises rapidly at first, when additional prey are easy to capture, but gradually flattens out as getting additional prey becomes increasingly difficult. The curve finally becomes flat when the maximum load size the forager can manage has been accumulated.

Central-place foraging decisions are related to the ease of capturing prey when some are already in the bill. Marsh-nesting blackbirds capture recently emerged prey that are still soft and unable to escape. In these

cases, capturing additional prey is evidently easy. I have seen blackbirds in Washington with over three dozen insects in their bills! Costa Rican redwings, however, have to pursue active prey. Apparently it is difficult for them to capture even a second prey while holding on to the first, so they usually return to the nest with one prey. Meadowlarks dig many of their prey from holes in the ground, and they gape for them in the bases of grass clumps. To capture subsequent prey requires dropping previously captured prey. Meadowlarks are nonetheless able to deliver multiple prey on each trip because they place previously captured prey on the ground while searching and then pick them up again as they move on. This is possible on dry, open ground but not over water where many blackbirds forage.

Another way of getting previously caught prey out of the way is to swallow them and regurgitate them later,

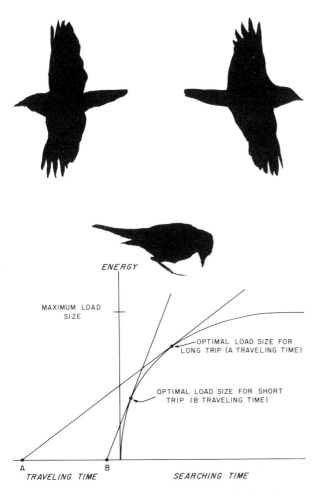

3.9 *Optimal load sizes for a bird bringing prey back to its nest. The graph assumes that the more food the bird already has in its bill, the harder it is to capture more prey. The best time to stop loading and return to the nest depends on how far away the nest is, because no food is captured during the round trip to and from the nest. If the nest is nearby, it pays to return with a relatively small load because little time is lost during the trip. If the nest is farther away it pays the bird to keep on loading even though its efficiency is dropping because it loses so much time during the round trip.*

or hold them in the throat where they do not encumber the bill. Many birds do the former but very few insectivorous birds store food in their throats. To the best of my knowledge no blackbirds are known to do so, but why this potentially useful method is avoided remains a mystery.

Blackbirds are good judges of the relationship between distance and load size and bring larger loads when they forage farther from their nests. I measured load sizes in relation to distances from nests among Brewer's blackbirds in the Potholes, where the vegetation on the edges of the lakes and uplands is short and foraging blackbirds are easily observed from the basaltic cliffs. Brewer's foraging within fifty meters

of their nests return quickly with two to four prey in their bills while those foraging more than one hundred meters from their nests return with fifteen to twenty or more prey per trip.

To minimize the amount of time they spend traveling between their nests and foraging areas we might expect that birds would place their nests as close as possible to the "center of gravity" of the food they are exploiting. Because nests are vulnerable to predation, however, and because the many trips to them by the adults may help predators find them, safety of nest sites may be much more important than their proximity to foraging areas. My studies of colonies of oropendolas and caciques in South America show that the birds moved a long way from the best foraging areas to gain safe, isolated trees in which to nest. At a colony of yellow-rumped caciques near Tingo María in Peru the birds did most of their foraging in the forest north of the nesting tree and very seldom arrived at the colony from the south (fig. 3.10). To the south there were mostly fields and pastures that provided little food for caciques. They could have shortened their flying distances by nesting in the forest, but the nests would have been more accessible to predators there.

How Blackbirds Learn What to Eat

In the normal course of events, young blackbirds receive a wide variety of prey from their parents and during the later part of their dependency period they are capable of seeing the prey clearly and perhaps forming associations between shapes seen in the bills of their parents and the prey they encounter later as they forage on their own. It is unlikely, however, that young birds could learn all they need to know about their food in this manner, because foods eaten change with place and season and the young soon encounter prey types never delivered to them by their parents. In a variable environment, flexible response patterns and curiosity are likely to be beneficial traits.

The role of curiosity is explored best by hand-raising nestlings on artificial diets that greatly limit their exposure to prey of different types, and then testing their responses to novel prey. John Alcock collected nestling redwings from the Potholes area and hand-reared them in Seattle, feeding them turkey starter mash (administered with an eyedropper), small pieces of water-soaked Purina puppy chow, and half mealworms. When the young were ready to leave the nest they were transferred to larger cages. They were still fed by hand with the same prey but, in addition, wild bird seed mixed with mealworms was available to them. About one week after they fledged the birds began gaping at newsprint on the cage floor and began to take mealworms on their own. At varying times

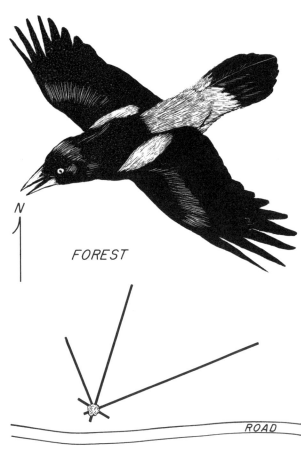

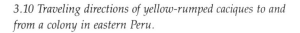

3.10 Traveling directions of yellow-rumped caciques to and from a colony in eastern Peru.

(two, six, ten, and twenty days) after taking their first mealworm the birds were deprived of food at seven P.M. and then tested between seven and eight the next morning. Each individual was offered a series of insects in succession: two mealworms, a frozen and defrosted damselfly, a freshly killed mealworm beetle, and then two live stinkbugs. The birds attacked twenty-six of the thirty-two damselflies and beetles offered to them, all in forty-five seconds or less, and they ate all but three of those attacked. The birds did not hesitate to attack the unfamiliar stinkbugs and most of them ate both the first and the second one even though they showed signs of finding the first one somewhat distasteful.

Thus, even though the young redwings had very limited exposure to prey types, they readily attacked insects that looked very different from those they had seen before and immediately ate them. In some sense they knew that these organisms represented potential prey the first time they encountered them. All of these prey are very different from mealworm larvae in size and appearance, making it unlikely that the birds were generalizing from their earlier experience with meal-

worms. If the stinkbugs really were highly distasteful the birds probably would have learned to reject them rather quickly, but the experiments were not continued to test this.

Redwings also learn a great deal about where prey are located and what kinds are expected in different places. John Alcock also tested this with hand-reared redwings using a maze in which prey were hidden in holes 0.75 centimeters in diameter and 1.5 centimeters deep in wooden planks of 10×2.5 centimeters. During a training period birds were presented with pieces of sunflower seeds or the rear half of a mealworm in sixteen of the fifty holes in either the lower or upper row of holes in the maze (fig. 3.11). The food items were placed in the holes selected at random with the constraint that no two adjacent holes had a food item. The items were placed so as to be flush with the opening of the hole. On each day a bird was allowed to forage for thirty minutes after it found its first bait. The birds were given the same bait distribution during the first two days; on the third day the same type of prey was available but the sixteen individual items were distributed evenly between the upper and lower rows. On the fourth day the birds were presented with sixteen items of food scattered randomly betwen the upper and lower rows, but this time eight were mealworm halves (four in each row of holes) and eight were sunflower seed bits (four in each row of holes).

The birds learned very quickly. On the first day it took them an average of twenty-four minutes to find the first food item, but on the second day they found their first prey after about three minutes and no bird took longer than nine minutes. Once they found the first item, they began searching for more prey. On the first day half of the birds removed ten of the sixteen prey within seven minutes of finding the first one; on the second day it took them about two minutes to find ten prey items. The improvement was largely owing to their learning to scan the holes from a distance rather than going up to each one and probing into it.

On the third day, when the distribution of the food changed from being in one row only to being evenly divided between the two rows, the birds at first searched preferentially in the rows where they had found food the previous two days, but they quickly learned that food was now available in both rows. The type of prey they had fed upon the first three days greatly influenced what they took on the fourth day when both prey types were available. Birds that had eaten only mealworms the previous three days took 62 percent to 100 percent of the larvae before they took their first seed. However, the birds that had been eating sunflower seeds switched much more quickly to taking mealworms. The majority of the first eight prey taken by these birds were mealworms. The larvae were larger,

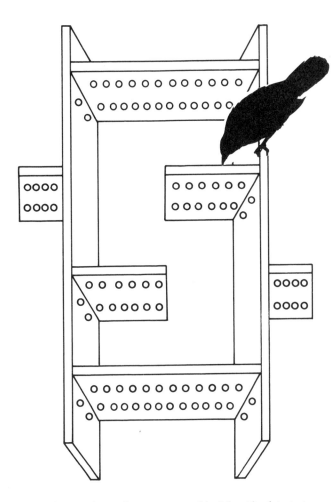

3.11 The experimental apparatus used by John Alcock to test how much foraging redwings learn about where prey are located and which kinds are found in which places.

easier to see, and probably more desirable as prey.

These experiments show that redwings are quick to learn where prey are expected, and the type expected there, and that they can rapidly change their behavior and expectations as prey availability changes. This is not surprising in a bird that encounters and eats such a wide variety of prey at different times and in many different habitats—often in quick succession. It also means that although they do not possess the perfect knowledge required of them by foraging theory, they probably know enough to base their choices on the criteria demanded of them by the theory.

How Blackbirds Take Care of Themselves While Also Feeding Young

The demands of feeding young are such that adult blackbirds often lose weight during the breeding season. Nonetheless, they must feed themselves enough to survive at least until their young are inde-pendent or else the effort is wasted. Also, if they kill themselves in their efforts to raise current offspring, they will not be able to raise any more in the future. But how much food do they reserve for themselves and how do they intersperse feeding themselves with caring for offspring?

During incubation a bird benefits by gathering food quickly so as to minimize time off the nest which may chill the eggs and expose them to predators. If both adults incubate, the typical pattern is for each sex to take one or a few very long bouts on the nest while the other forages. But in all blackbirds for which we have any information, only the female incubates. Therefore, the eggs are unattended while the female is absent and the pattern of incubation is different. Many short feeding trips are interspersed with longer incubation sessions during the day. Time off can be reduced if the male feeds the female, but this is very rare among blackbirds, having been recorded only in the melodious blackbird.

When young are being fed the adults forage most of the day and are able to feed themselves regularly. However, many different patterns of self-feeding are possible. For example, the adults could fill themselves up at regular intervals and then concentrate on delivery of food to the young the rest of the time. Alternatively, the adults could feed themselves a little on each trip before they started loading for their offspring. In the few species for which we have any data, the pattern appears to be closer to the latter. In the Potholes, female redwings and yellowheads fed themselves briefly on about 79 percent of all foraging trips I observed—the self-foraging times ranging from about one to three minutes in duration. During those times they ate the same types of prey they subsequently loaded for their young with one striking exception. Whenever an adult captured a large dragonfly, it always stopped feeding itself and immediately began accumulating a load. Evidently the value of such large prey for delivery to a distant nest is so great that they are never eaten. On the other hand, adults foraging in the same places but without dependent young regularly ate the dragonflies they captured.

Do Blackbirds Affect Their Own Food Supplies?

Even though blackbirds regularly starve, it does not follow that they are having any major impacts on their food supplies. Not surprisingly, the best information available concerns the impacts of blackbirds on crops. As early as 1667 Massachusetts Bay colonists enacted laws to attempt to reduce blackbird damage to their corn crops. One such law even specified that "every unmarried man in the township shall kill six blackbirds

. . . [and] as a penalty for not doing it, shall not be married until he obeys this order." We don't know if this law doomed anyone to bachelorhood, but it probably did little to limit the growth of blackbird populations.

Farther west, pioneers in the Great Lakes regions, as well as the Indians before them, had similar blackbird problems. In 1749 a French explorer, J. G. C. De Lery, noted that blackbirds were so abundant around western Lake Erie that people were assigned to guard the ripening crops. The Lake Erie region continues to be one where blackbird problems are more severe than in most parts of the country.

Redwings attack maturing corn ears when they are in the milk stage of development, using gaping movements to slit the husks and expose the kernels. Fully mature, hard corn is relatively immune to redwing attack because the birds cannot handle the hard kernels except with difficulty; but grackles, with their more powerful bills, crack mature kernels easily. In the western Lake Erie region most of the birds causing damage to corn fields nested or were raised in nearby counties. Banding recovery data indicate that redwings move no further than thirty-six miles on the average between their nesting and roosting localities.

Despite the attention given to blackbird damage to corn crops, the average annual loss to Ohio corn growers has been less than 1 percent of the total crop. This amounts, however, to 2.5 to 3.0 million bushels at a value of 3.8 to 6.8 million dollars annually. The real reason why there is so much concern, however, is not the average loss rate but the fact that those farmers close to roosts receive very high losses. Of 7,237 fields examined in nineteen Ohio counties during 1968-76, 85 percent received less than 1 percent damage and only 2.5 percent of fields received greater than 5 percent, the level at which investment in damage protection is normally economically worthwhile. Three counties bordering western Lake Erie contained 62 percent of the fields where losses exceeded 5 percent, and 77 percent of those where losses exceeded 10 percent. Nearly all fields with losses greater than 5 percent were within five miles of a major marsh roost of blackbirds (fig. 3.12), and the closer the field to the roost the greater the level of damage (fig. 3.13).

The reasons for this pattern of damage are not difficult to determine. Blackbirds at this time of the year are making daily foraging trips from their roosting marsh to the corn fields. If good foraging is available close to the roost it does not pay to travel farther because additional travel merely wastes time and energy. What is perhaps more puzzling is why any birds fly more than a few miles, because even in the closest fields damage levels are seldom as high as 15 percent. The answer may be that corn is available for

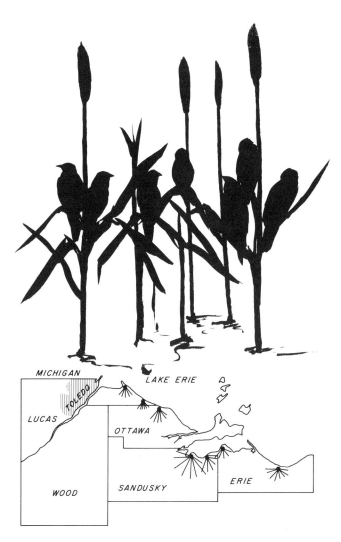

3.12 Locations of major roosts of blackbirds in late summer in marshes at the western end of Lake Erie.

easy harvest for only a short time in any one field because ears become too hard before the birds have been able to damage more than a modest fraction.

It is more difficult to determine the effects of blackbirds on their insect prey. Blackbirds are more common on productive than nonproductive marshes and there is some nestling starvation even on productive lakes, but the birds may not seriously reduce the levels of their prey. Starvation is often associated with periods of bad weather during which it is difficult for the adults to forage without exposing the young to chilling temperatures. Also, emergences of aquatic insects are low on rainy days, causing a temporary shortage of food that is not owing to overexploitation.

My best estimates of the effects of blackbirds on insect populations were made at Coot Lake, a small lake in the Potholes, where I operated five emergence traps to sample aquatic insects as they left the water and metamorphosed into terrestrial adults. Coot Lake was

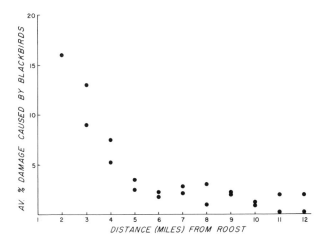

3.13 *The amount of damage caused by blackbirds to corn crops decreases rapidly with distance from roosts near Lake Erie in western Ohio.*

convenient because it lacked cattails and bulrushes in the water—the entire edge being composed of short grasses and sedges. In addition, from nearby adjacent basaltic cliffs I could see all of the blackbirds foraging at the shore of the lake. To determine the diurnal pattern of emergence of insects, I visited the lake every two hours on a number of days scattered throughout the breeding season and counted both the blackbirds and insects on each visit. By combining this information with other data on the rate at which blackbirds captured insects while foraging at the lake's edge, and knowing what fraction of the lake edge my traps covered, I estimated the proportion of emerging insects captured by the birds.

From my trapping I estimated that about 23,500 odonates (dragonflies and damselflies) emerged from Coot Lake every day. With an emergence period of about 125 days (early May to early September), the total emergence would have been about three million insects. During the day there were, on the average, six blackbirds present capturing prey at the rate of about fourteen per minute. Of those prey, about 80 percent were probably odonates. Thus the birds were capturing about 55,000 odonates per day—about twice the estimated emergence! Clearly this is impossible, but even with reasonable corrections it is difficult to reduce the estimated number below the calculated emergence. Obviously, birds were capturing a large fraction of the odonates emerging from Coot Lake, perhaps as high as 95 percent of them. Even so, blackbirds may have little effect on the sizes of odonate populations because each damselfly lays about fifteen hundred eggs in her lifetime. Only two thousand females are needed to lay three million eggs, the minimum number required to yield the estimated emergence. Not all females that emerge live long enough to lay eggs, and there is high mortality among larvae in the water, but only a tiny fraction of emerging females, perhaps less than 1 percent, would need to survive blackbird predation to maintain the odonate population at the level that can be supported by the lake.

What would happen to aquatic insects if there weren't any blackbirds? There might not be any more of them, but they might evolve different behavior. The strongly synchronized emergences of damselfies during the morning is almost certainly a response to heavy predation, and if that were relaxed emergence times might be more spread out. Nonetheless, emerging aquatic insects exert a much more powerful influence on blackbirds than *vice versa*. We know even less how arboreal icterids affect populations of their prey. Given the complexities of those environments compared to the relatively simple marshes, we may never know.

4.1 During boundary disputes male redwings use the typical,
species-specific bill-up posture (left) while male yellowheads
use the typical asymmetrical song spread with the head turned
over the left shoulder (right).

4. Food and Blackbird Distribution

In nature resources are often so abundant that birds can meet their needs in a relatively short period of time. At other times food is scarce enough that birds may not find enough even if they forage all day. Such shortages need not come yearly to have an important influence on the distributions of species, their overall abundances, and the evolution of their use of environmental resources.

The distributions and abundances of blackbirds are clearly influenced by food supplies, but determining when food shortages occur turns out to be more difficult than might have been imagined. Starving birds are seldom found in nature, probably because when an individual becomes weakened it dies very quickly. Also, an individual having difficulties finding food may take more risks while foraging and be more likely to become a meal for one of its own predators. It is relatively easy to correlate population densities with estimates of food abundances and to note reductions in bird populations during times of probable resource shortages, but these measures are indirect. They do not tell us exactly what is happening. Nonetheless, slow progress is being made in understanding the nature of food shortages, how animals compete for scarce resources, and the results of competition. Blackbirds have contributed in a modest way to our understanding of these processes.

How Birds Compete

Competition for food occurs when resources are insufficient to meet demands, and the harvesting of resources by one individual adversely influences the ability of other individuals to harvest the same resources. Such lowering of feeding rates may happen in several ways.

First, by harvesting some prey items a forager may leave fewer prey for another individual to find. This is called resource *depletion*. The immediate effect is that another individual (or the same individual at a later time) foraging through the same environment experiences a lower rate of encounter with prey than it would have if the first individual had not hunted there.

Alternatively, the passage of one individual may change the behavior and locations of prey so that another individual passing through the same place a bit later finds fewer prey even if the first predator did not actually capture any of them. This is known as prey *depression*. For example, if a hawk makes an unsuccessful attack on a flock of birds, all members of the flock are likely to be highly alert for predators for some time afterwards. The chances of a successful attack during that period are probably lower than they would have been if the first attack had not been made. Although the immediate effect of prey depression is the same as depletion, there is an important difference because prey populations usually recover faster from depression than from depletion. Recovery from depletion requires either the birth of new individuals or immigration into the area, while recovery from depression requires only a change in behavior.

The third way in which one predator can influence the ability of another to forage in an area is by direct physical *interference*. One individual may displace another from a foraging site or defend a large enough area that the individual can meet all of its needs for a breeding season or lifetime within its confines. These behaviors can prevent other individuals from using the resources of those areas for long time periods.

Under some circumstances the passage of one predator may actually *enhance* the ability of another predator to find prey. For example, the first predator may cause prey to move to a location where they are more easily found by a second predator. This is most likely to occur if the two predators belong to different species and have different modes of seeking prey. Some birds associate in flocks containing several species because individuals of one species enhance prey availability for individuals of the others. Having more eyes on the lookout for predators is also an advantage to all individuals in a flock, whether or not they belong to the same species.

The immediate consequences of reduced prey availability is a lowered prey encounter rate, but the long-term consequences may be quite different. A competitor may actually prevent individuals of other species from finding enough food to reproduce so that the species eventually disappears. Such competitively induced extinction has probably occurred many times during evolutionary history, but it is extremely difficult to document. Past extinctions leave few traces of their causes. There is still a great deal of controversy over whether dinosaurs went extinct because of competition with newly evolving mammals or because of adverse climatic changes, or both. Most current extinctions are caused by direct hunting or habitat destruction, as human activities are fostering what may be the highest level of extinction ever recorded in the history of life. But this may not help us understand past extinctions.

A common and more readily documented effect of competition is that a competitor is kept out of habitats it would otherwise occupy. One reason why habitat exclusion is such a common result of competition follows directly from foraging theory. If two competitors have similar but not identical food and habitat preferences, the foraging activities of one reduce food levels in some habitat patches more than others. This causes some patches to fall below the level at which it pays individuals of the other species to forage in it, as shown by curve A in Figure 3.6. When this happens, the second species no longer forages there. But enough of theory. Let us look at how blackbirds compete and influence one another's distribution patterns.

Competition and Blackbird Distribution Patterns

The most conspicuous way in which blackbirds influence each other's distribution is through active defense of breeding areas. Many blackbirds are territorial, but most of them defend their territories only against other individuals of the same species. Individuals of one species, however, are sometimes prevented by individuals of another species from breeding in areas that would otherwise be quite suitable.

The best-known case of interspecific territoriality among blackbirds is that between breeding red-winged and yellow-headed blackbirds (fig. 4.1). Of the two species, the redwing has the wider breeding range and it occurs in nearly every marsh where yellowheads breed. The yellowhead is larger, males weighing about 100 to 105 grams, while redwing males weigh about 65 to 70 grams. This size difference is sufficient to give yellowheads a substantial advantage in contests over territories. When yellowheads arrive on their breeding marshes, they usually evict already established male redwings from their territories (figs. 4.2, 4.3). But there are many more male yellowheads than actually establish territories. Many of them become floaters rather than evicting male redwings from their territories. Why they do not do so is a puzzle, some of the pieces of which are beginning to emerge.

Yellowheads do not forage easily in trees and are inhibited by the presence of trees adjacent to their breeding marshes. I was first alerted to this fact when watching blackbirds in the Cariboo Parklands of British Columbia where most of the marshes are surrounded by open grasslands and have nothing but yellowhead territories on them. Woodlots border the marshes in some places, and where they do I found redwing

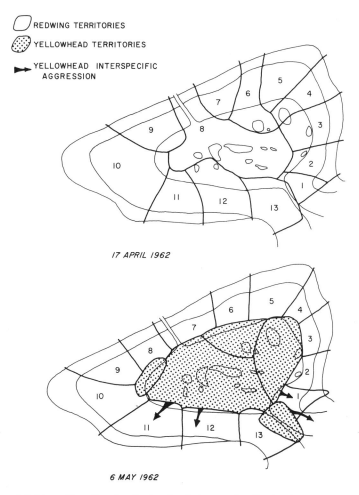

REDWING TERRITORIES

YELLOWHEAD TERRITORIES

YELLOWHEAD INTERSPECIFIC AGGRESSION

17 APRIL 1962

6 MAY 1962

4.2 *Map of breeding territories of redwings and yellowheads on MacDowell Lake, Turnbull National Wildlife Refuge. On 17 April, before the arrival of yellowheads, territories of male redwings occupied most of the lake. In early May, male yellowheads evicted the redwings from the center of the marsh and five yellowhead territories were carved out of that space.*

territories in those parts of the marshes closest to the trees. While making measurements of those marshes I noticed a curious thing. If the angle to which the tops of the trees projected above the marsh was less than 30 degrees, yellowheads were present; but if trees projected more than 30 degrees, yellowheads were absent. I later discovered that the same pattern held in Washington (fig. 4.4). In the Potholes, cliffs produced the same effect as trees did elsewhere. Cliffs do not actually interfere with upland foraging opportunities the way trees do, but cliffs are found near a marsh so rarely in western North America that yellowheads have not evolved mechanisms to discriminate between trees and cliffs.

This interesting pattern suggests that whether yellowheads displace redwings depends not only upon the dominance given them by their relative size, but

4.3 *A marsh at the Turnbull Refuge. Solid lines indicate redwing territories; yellowhead territories are stippled.*

also upon the vigor with which territories are defended. It is impossible to ask yellowheads how much they want to occupy a site, but their behavior suggests that they do not try as hard to defend places adjacent to trees and cliffs. Since redwings forage readily in trees, they are evidently not inhibited by them. Thus, the vigor of their defense of space is undiminished near tall trees, perhaps giving them the edge over the larger but somewhat less motivated yellowheads in those circumstances.

Redwings also have strong interactions with another very closely related marsh-nesting blackbird in California—the tricolored blackbird. The tricolors are restricted to the lowland valleys of California, with the exception of a small breeding population in the Klamath area of southeastern Oregon. Redwings are also present throughout this range, and both species are intimately associated during the breeding season. The tricolors are densely colonial and the territories of individual males are no more than one to three meters on a side. In larger colonies over one hundred

thousand birds may be present. Not all birds begin nesting simultaneously, but synchrony is great enough that most nests in the colony have eggs and young at the same time.

Male redwings frequent their breeding areas in lowland California all year, and they begin to defend their territories during early morning and late evening hours by mid-winter. At that time tricolors are still traveling in large flocks and visit marshes only at night to roost. By the time tricolors are ready to breed, redwings are fully established on their territories and many nests have already been built. Tricolors arrive at their prospective breeding sites *en masse*, and hundreds may settle on a single redwing territory. When this happens the male redwing responds with a frantic effort to evict the mob, but even though they offer no resistance to his attacks, simply moving a few feet each time, he is quickly exhausted and within a few days the issue is settled—the tricolors nest where they wish (fig. 4.5)! Any redwing nests already underway at the time of settlement are abandoned and the females move

4.4 On lakes that are otherwise suitable for yellowheads, territories are not established in places where trees and other objects project above the horizon at angles of greater than about 30 degrees. This illustration shows a yellowhead territory in the deeper water, where the trees subtend angles less than 30 degrees with the horizontal, and a redwing territory closer to shore where the trees project higher than that.

elsewhere. However, in contrast to the case of competition with yellowheads, where many of the best breeding marshes in a major part of the breeding range of the yellowhead are made unavailable to redwings, the area occupied by the dense colonies of tricolors is so small that the overall effect on redwing populations is insignificant.

Interspecific territoriality is well developed among the two species of meadowlarks in North America. Eastern and western meadowlarks overlap as breeding birds in an extensive latitudinal belt from Minnesota and Wisconsin south to the southern Plains. In this belt easterns favor wetter habitats, especially along rivers, while westerns prefer drier ridges and hills. In many places, however, both species can be found breeding together. In Wisconsin and Kansas, where they have been studied intensively, the species are strongly interspecifically territorial—their territories are completely segregated even when intermixed. The two meadowlark species are extremely similar in appearance, but their songs are strikingly different. By playing songs to territorial males in areas where both species were breeding and in areas where only one species was present, Sievert Rohwer demonstrated that males respond with territorial behavior much more strongly

to songs of their own species than to songs of the other. In areas where both species occurred together, however, the songs of the other species elicited stronger responses than they did in areas where the birds did not normally hear those songs and may not have been familiar with them. Their lack of response suggests that they do not innately recognize songs of the other species as territorial threats.

These results suggest that song is used less by meadowlarks in defense of interspecific territories than in defense of territories against males of the same species. There is more fighting between than within species, perhaps because singing is not effective at averting fights when males sing different songs. In areas where both species were present, 30 percent of males collected had been wounded in the head region, while in areas where only one species bred, only 15 percent of males had been wounded. Wounds were also more common in the rarer of the two species in each locality, 26 percent of members of the more common species being wounded compared to 35 percent of members of the rarer species. In addition, male western meadowlarks, the smaller of the two species, were significantly larger in areas where they bred in proximity to eastern meadowlarks than they

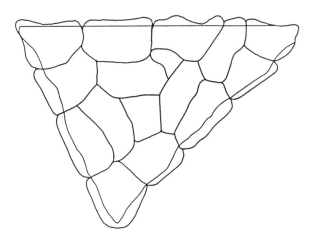

15 MARCH 1959

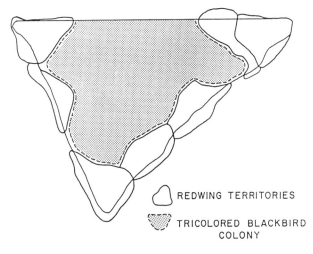

20 MARCH 1959

REDWING TERRITORIES

TRICOLORED BLACKBIRD
COLONY

4.5 *This small marsh near Los Angeles held sixteen redwing territories on 15 March. A few days later a large flock of tricolored blackbirds invaded the area and began nest building, evicting most of the redwings.*

were in areas where they bred alone. In other words, having big competitive neighbors favors being large yourself.

These results are interesting, but the meadowlark story is even more complex. In the southern Great Plains, eastern meadowlarks extend westward as breeding birds along the Canadian, Red, and Brazos rivers of western Texas, but they stop where the flood plains become narrow and the rivers are confined to rugged canyons. Farther west, after a gap of eighty miles or more, another race of the eastern meadowlark, Lilian's meadowlark, breeds in the dry-desert grasslands of western Texas, New Mexico, and Arizona, south into extreme northern Mexico. There is then

another gap until one encounters still another race of eastern meadowlarks farther south in Mexico. The western meadowlark breeds throughout this area and over the entire range of Lilian's meadowlark as well. Surprisingly, the ecological relationships of the two species in the west are the reverse of what they are in the east. In the southwest, the western meadowlark occupies the wetter sites, whereas Lilian's breed in the very arid desert grasslands. In the Plains and Midwest, however, westerns occupy the drier sites and easterns the wetter ones. Everywhere, the two meadowlarks defend territories against one another when their breeding areas overlap. Lilian's meadowlarks are as different in shape from the eastern race of the eastern meadowlark as either is from western meadowlarks, and they may actually belong to a third species of meadowlark. Since a particular site can accommodate no more than one kind of meadowlark, there is intense competition between species, which leads to both habitat segregation and changes in shapes and sizes of the birds.

In South America, meadowlarks all have red rather than yellow breasts and they, too, are sometimes interspecifically territorial. Greater and lesser red-breasted meadowlarks overlap in their breeding distributions in southern Buenos Aires Province in Argentina. In those areas males defend their territories against males of both species, the lesser usually by a flight display, the greater while perched.

The direct, easily observed fighting among redwings, yellowheads, and meadowlarks leaves no doubt that they restrict one another's ranges. Distribution patterns suggest that similar interactions are occurring among other species, even though evidence is not yet available.

The genus *Agelaius* probably evolved in South America, where most of the species occur today. The ancestors of today's redwings invaded North America at some unknown time in the past and spread over the continent. During the recent isolation of the California lowlands from the rest of the continent during the Pleistocene glaciations, the tricolored blackbird evidently evolved from the local redwing stock. Two different invasions of the West Indies have also yielded island species of the genus. Nonetheless, over most of North America the redwing is the only *Agelaius* species, and it currently breeds from southeastern Alaska to the tropical marshes of Costa Rica.

Among the marsh-nesting blackbirds in South America there are several widespread species that replace one another geographically, plus some rare species with limited distributions. The temperate marshes of South America are occupied by the yellow-winged blackbird. It is replaced rather abruptly in northern Argentina and southern Brazil by the chestnut-capped blackbird, which extends north to

TROUPIAL

EPAULET ORIOLE

MORICHE ORIOLE

YELLOW ORIOLE

YELLOW-TAILED ORIOLE

YELLOW-BACKED ORIOLE

WHITE-EDGED ORIOLE

ORANGE-CROWNED ORIOLE

4.6 *The orioles of South America and their ranges.*

French Guiana and eastern and central Brazil. The tropical marshes of northern South America are occupied by the yellow-hooded blackbird, which also occupies the temperate marshes of the Andes in Colombia. These three species, which are often common where they do occur, are largely exclusive in their breeding distributions. In addition, there are several rather local and much less common *Agelaius* species. The unicolored blackbird occurs as an isolated population in northeastern Brazil north of the Amazon but does not appear again until the marshes of coastal Río de Janeiro and the drainage of the Río Paraná. The center of its distribution appears to be the Paraná valley in Bolivia, Paraguay, and northern Argentina. Little is known of its habits. The pale-eyed blackbird is known only from lowland marshes, principally those found along the margins of oxbow lakes, in eastern Peru. It was discovered only a little more than a decade ago! We can only guess about the factors responsible for these patterns. History doubtless has something to do with determining where the boundaries are, but competition may be the cause of the persistence of those boundaries.

Restricted ranges are also characteristic of the all-black upland blackbirds of South America, which are members of several genera. The austral blackbird is found only in extreme southern Argentina and central and southern Chile. Forbe's blackbird, a rare and little-known species, is recorded only from Pernambuco and Minas Gerais in Brazil. The Bolivian blackbird occurs only in a few dry montane valleys on the east side of the Andes in Bolivia. The golden-tufted grackle is found in Guyana and immediately adjacent Venezuela and Brazil while its close relative, the mountain grackle, is found only in a restricted part of the east slope of the Andes in Colombia.

The most widely distributed of the all-black South American blackbirds are the velvet-fronted blackbird, which occurs along rivers in the entire Amazon Basin, and the chopi, which ranges from east and central Brazil south to northern Argentina. Nonetheless, most areas in South America have only one common all-black blackbird. Most of the range boundaries are not obviously related to the presence of a potential competitor, but most boundaries occur in very poorly-known areas. Many interactions that we have yet to detect may be occurring there.

Oriole species are also surprisingly restricted in their ranges in South America (fig. 4.6). In the entire continent there are only eight species of breeding orioles as compared to fourteen in Mexico alone. Of these species, only one occurs as far south as Argentina even though northern Argentina extends well into the subtropics. Only two have ranges extending south of the Amazon River. In contrast, caciques are common

and widespread in South America. The yellow-rumped cacique is found throughout northern South America and the Amazon Basin while the red-rumped cacique occurs still farther south, although with many gaps in its range, to northern Argentina and Paraguay. The similar scarlet-rumped cacique ranges in montane forests south along the Andes to northern Peru. The golden-winged cacique is widely distributed from southern Brazil and eastern Bolivia south to northern and central Argentina. The similar selva cacique is confined to eastern Peru. There are also four species of all-black caciques whose combined distributions cover the forest zones of tropical and lower montane South America to northern Argentina.

Orioles are notably scarce in the areas where caciques are most common. That the distributions and abundances of orioles and caciques may be mutually related is suggested by the pattern in Central America. The Mexican cacique has a very limited range in western Mexico, while the only other cacique found north of Honduras, the yellow-billed cacique, is a bird of dense forest understory. Over most of the area where orioles are common in Middle America, caciques are rare or absent. Farther south where caciques become more common, as in eastern Costa Rica where the scarlet-rumped cacique is a conspicuous bird of the wet forests, orioles are rare and local as they are in much of South America.

There is no evidence of direct behavioral interactions between orioles and caciques, and the inverse relationship among their abundances may be caused by other factors. For example, orioles are more common in arid and semiarid habitats whereas caciques are more common in humid areas. This is not likely to be the complete answer, however, because there are humid region orioles and the Mexican cacique occurs in very arid country. Cacique-oriole relationships need much more study.

Overlap in Resource Utilization by Blackbirds

Competition is rather difficult to demonstrate directly, but it is relatively easy to determine the extent to which different species overlap in their use of environmental resources. Again it is the marsh-nesting species of North America that are the best known. The yellow-headed blackbird, which occurs in the western and central parts of North America where marshes are highly productive, is a specialist on aquatic prey. Yellowheads mostly bring aquatic insects to their nestlings, but adults do take advantage of crop harvesting in the vicinity of their nests to bring butterfly and moth larvae to their young.

Redwings also make extensive use of marsh insects

Table 4.1. **Foods Delivered to Nestling Redwings and Yellowheads at Rush Lake, Iowa, in 1969 (expressed as percent of total volume of prey)**

Prey Type	Aquatic or Terrestrial	May 22–June 6		June 7–June 20		June 21–July 4	
		Blackbird Species					
		RW	YH	RW	YH	RW	YH
		Nests Sampled					
		8	14	14	38	28	14
Damselflies	A		25.0	6.4	20.1	30.5	
Crickets	T			7.3			
Leaf bugs	T					5.4	
Treehoppers and aphids	T	5.1		5.1	3.1	8.1	
Beetles	T	20.5	9.4	10.4		12.6	
Moths	T	47.9	11.0	35.3	8.9	14.0	
Aquatic flies	A		30.9	12.8	41.3	10.4	37.7
Terrestrial flies	T		3.0				23.1
Spiders	Both	6.8	3.8	3.8	3.9	9.8	4.3
Snails	A	6.0					
Unidentified	?	5.1					
Corn seeds	T						3.0

Note: Table adapted from Voigts 1973. Only prey comprising more than 3 percent of total are shown.

but they forage more in uplands around the marshes than do yellowheads. These differences in foraging show up clearly in the food delivered to the nestlings. In Iowa, David Voigts found that yellowheads brought more damselflies and aquatic flies to their young than did redwings, while redwings brought more moth larvae (table 4.1). Measured by volumes of prey delivered, there is an overlap of about 50 percent in the types of prey delivered to the nestlings.

The same is true in the Potholes of eastern Washington where redwings and yellowheads bring similar food to their young (table 4.2). They also capture most of their prey in the same types of places. In fact, redwings and yellowheads overlap 65 percent in the places in which they forage, and 85 percent in the types of prey they deliver to their young. The biggest difference between the two species is that the larger yellowheads seldom forage within the dense canopies of the sagebrush bushes while the smaller female redwings regularly do. As a consequence, the redwings bring many small caterpillars to their nestlings while yellowheads do not. However, female redwings capture and bring to their young prey as large as the largest delivered by female yellowheads to their young. The great degree of overlap in resource utilization is probably why redwings and yellowheads defend

territories against one another. The presence of foraging redwings within the territories of yellowheads would doubtless reduce the encounter rates with prey for yellowheads, and *vice versa*.

While marshes are very productive biologically, they are structurally simple environments. They have only a few types of prey in them and they offer only a few ways by which a bird can search for prey. Therefore, we expect any two blackbirds nesting in the same marsh to overlap a great deal in the food they bring to their young if they both forage in the marshes as do redwings and yellowheads. Redwings and common grackles breed together in some marshes, as at the Wingra Marsh in Madison, Wisconsin, but their diets are quite different (table 4.3). The grackles forage much more in adjacent uplands for their food and mostly bring moth larvae to their young, while redwings use the marshes more.

In Argentina I was surprised to find that the three common marsh-breeding blackbirds brought very different foods to their nestlings. The most important prey delivered to nestling brown-and-yellow marsh-birds were crickets, lepidopteran larvae, and spiders. Nestling yellow-winged blackbirds were mostly fed damselflies, dragonflies, small flies, and spiders, while nestling scarlet-headed blackbirds received a fare of

Table 4.2. **Food Delivered to Nestling Blackbirds in the Potholes, Columbia National Wildlife Refuge, Washington (expressed as percent of daily total of both prey items and calories)**

Prey Type	Prey Items			Calories		
	Yellowhead	Redwing	Brewer's	Yellowhead	Redwing	Brewer's
Dragonflies and damselflies	56	37	34	64	35	34
Grasshoppers	1	2	7	3	7	24
Cicadas	1	3	1	4	23	11
Butterflies and moths	4	30	24	3	14	13
Flies	24	13	21	6	5	5
Other insects	7	5	7	4	3	3
Spiders	4	7	5	10	9	9
Other items	3	3	1	6	4	1
Total items	1,417	1,901	1,070			

grasshoppers, beetles, moth larvae, and spiders. Nestling diets overlapped only 15 percent between brown-and-yellows and yellow-wings, 19 percent between brown-and-yellows and scarletheads, and 26 percent between scarletheads and yellow-wings.

Why are there such marked dietary differences among these blackbirds when their North American counterparts are so similar to each other? Even though the Argentina marshes are structurally very similar to North American marshes, they support much lower populations of insects. My sampling efforts suggest that emergence rates of aquatic insects are less than 5 percent of the rates I found in Washington. As a result, only one of the species, the yellow-wing, forages extensively on emerging aquatic insects. The scarlet-head uses its sharp, chisel-like bill to hammer into and split open the stalks of cattails and bulrushes from which it extracts most of its prey. The pathway of a foraging scarlethead, even when the bird is continually out of sight, is signaled by the sharp sounds of its bill hammering on the stalks of emergent vegetation. The brown-and-yellows generally dig in the turf at the wet edges of the marshes.

These results also point to a problem encountered in attempting to understand the distribution patterns of birds, and the role competition plays in determining

them. It would be nice if one could tell if an area had as many species in it as the resources could support by examining the amount of overlap in their diets. Some theorists have developed models to predict theoretical limits to the amount of overlap that should be possible among species competing for food. Redwings, yellow-heads, and Brewer's are near those theoretical limits and North American marshes may, in fact, be "saturated" with species. The Argentine species fall well below those limits, but this does not mean that more blackbirds could be accommodated in those marshes. There are other marsh-nesting blackbird species a bit farther north and occasionally one of them, the chestnut-capped blackbird, breeds as far south as the area where I made my studies, but it does not establish regular breeding populations there. The amount of overlap that is compatible with coexistence may depend on overall resource abundances and how they vary. There is, unfortunately, no simple overlap measure that can tell us how blackbirds influence one another's distributions.

Table 4.3. **Prey Delivered to Nestling Redwings and Grackles at East Wingra Marsh, Madison, Wisconsin, in 1965 (expressed as percent of total volume of prey)**

Prey Type	20 May–10 June Redwing	Grackle	11 June–30 July Redwing
Earthworms		9.1	
Dragonflies and damselflies	13.9	5.3	8.7
Grasshoppers			43.2
Beetles		27.2	
Butterflies			8.6
Moths	73.7	17.4	10.2
Flies	3.7		6.9
Fulgorids		3.6	
Fish		7.0	
Bread		12.8	

Note: Adapted from Snelling 1968. Only prey comprising more than 3 percent of total are shown.

How Many Species of Blackbirds Live Together?

Suppose you were planning a bird-watching holiday, and wanted to go where you could find the largest number of species of blackbirds. Where should you go? Depending on your travel budget (you need to know Spanish in either case), I would suggest one of two places. The first would be the southern end of the Mexican Plateau and adjacent lowlands, where you could find twenty-four species, thirteen of them orioles. Farther north or south, prospects are not as good. Costa Rica, so rich in birds in general, has only fourteen breeding species of icterids and all of the United States only nineteen species.

The second favored spot is the Andes and adjacent lowlands of Colombia, where twenty-seven breeding species can be found. You would find only five oriole species there, but in their stead a richness of oropendolas and caciques (twelve species) plus some other northern Andean specialists. Both regions have great variability in climate and topographic relief. The large numbers of species is possible in part because of this environmental variability, but this is not all of the story.

At higher latitudes, the number of icterid species rises again in two specific areas (fig. 4.7). One of these is northeastern Argentina and adjacent Paraguay, Uruguay, and Brazil, where diligent search should net you nineteen species. The other is the American Mid-

west, where ten species breed. Neither of these is a region of great topographic or climatic diversity. In fact, they are both very flat, breadbaskets of their respective continents. But they have lots of marshes, and blackbirds love wetlands. Interestingly, more species of blackbirds breed in temperate marshes than in tropical ones, but the reverse is true for forests. Most temperate forests have only one icterid in them (an oriole) and some lack icterids entirely. There are none in the North American coniferous forest belt except in nonforest vegetation such as marshes and bogs. Orioles breed in Central American pine forests, but not in pine forests north of Mexico. Tropical pine forests have good growths of epiphytes, particularly bromeliads, that are lacking in temperate pine forests; and it is likely that bromeliads permit the survival of orioles there. In Belize and Honduras, yellow-backed orioles regularly probe and gape in bromeliads on the pine trunks and branches. The endemic Jamaican blackbird also forages mostly in bromeliads and other epiphytes in montane forests. The critical, but impractical, experiment would be to remove bromeliads from a section of pine forests to determine if resident orioles leave.

There are always more species of breeding birds in structurally complex habitats than in simple habitats. A typical grassland supports perhaps a half dozen species of breeding birds while a nearby forest may harbor several dozen. Forests are not necessarily more productive than grasslands, but there are more prey types and, even more importantly, more ways in which to hunt for them than are possible in a grassland. In forests there are peckers into bark, gleaners of bark, gleaners of small branches, gleaners of leaves, salliers, hoverers, and sit-and-wait predators. There is no way to do most of those things in a grassland or a marsh.

Blackbirds, however, do not increase in variety in structurally complex environments. The number of species in rich marshes and grasslands is about the same as in forests. In the temperate zone the number is actually greater. In northern Argentina there are seven marsh-breeding blackbirds. Not all breed in the same marshes, but they can be found within a single county. Blackbirds also like grasslands. In the Midwest it is easy to hear singing males of bobolinks and both species of meadowlarks in the same field, plus the calls of cowbirds seeking nests in which to lay their eggs. The Argentine pampas will similarly provide sounds of three species of meadowlarks and a cowbird.

Extensive deciduous forests cover eastern North America and there are outliers elsewhere on the continent. Baltimore and orchard orioles occur in most of those forests, especially around the edges and in disturbed places. At equivalent latitudes in South America there is almost no land, but the austral blackbird does occur in south temperate forests, both

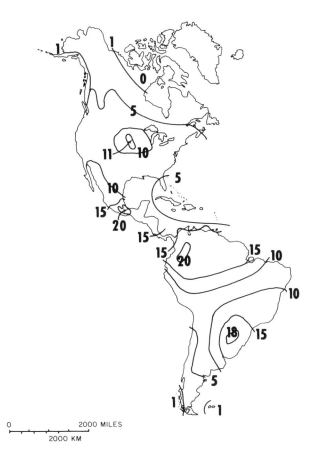

2000 MILES

2000 KM

4.7 Numbers of species of blackbirds that breed in North and South America.

coniferous and broad-leaved. In the southern beech forests they move around in small groups, acting very much like jays, which are absent from South America south of the tropics. Austral blackbirds also live in the mountains of Chile, where they forage in the giant araucarias, often called monkey puzzle trees in the United States. The branches of these remarkable trees are covered over much of their length by triangular, heavy, tough leaves that provide all sorts of hiding places for small animals. You have probably already guessed what the austral blackbirds were doing—they were gaping into those leaves, extracting spiders and insects from the crevices.

Icterids are also fond of deserts. The melodious songs of Scott's orioles enliven the dry yucca flats of the southwestern United States and Mexico. The thorny woodlands of northwest Mexico are home to Mexican caciques, red-eyed cowbirds, and two species of orioles. The very arid northern part of the Yucatan Peninsula has three species of orioles and the melodious blackbird. The Caribbean coast of northern South America is an arid pocket surrounded by humid lands. There troupials and yellow orioles are among the most conspicuous birds. Scrub blackbirds and white-edged orioles will reward visitors to the arid coastal deserts of southwest Ecuador and adjacent Peru. In all of these areas, wherever there is a reasonable growth of grass, some meadowlark, yellow- or red-breasted, can be found digging prey out of the ground and the bases of grass clumps.

So if you enjoy watching blackbirds and don't care how many species of blackbirds you find, you can go to most habitats. But I am fussy and, for reasons known to nobody, including myself, I like to have my feet wet. I have mostly slogged through marshes where interactions among blackbirds are diagrammatic and easy to see, the birds are abundant, and I don't have to climb trees to find their nests.

5.1 Great-tailed grackles gathering in a leafless oak tree. The two large males are readily distinguished from the females.

5. Gains from Grouping

One of my favorite childhood recollections is of watching large flocks of red-winged blackbirds pouring into their fall and winter roosts at dusk. Thousands of birds settled into dense cattail patches where they raised loud choruses of songs and calls until the fading light silenced them. Their dawn departure was equally impressive, but I seldom arose early enough to watch them then.

In spring I noticed that these large flocks gradually dispersed. Males began to defend territories in the marshes to which females were attracted. Small flocks of roving birds came through the marshes, seeking unoccupied territories, but the large fall and winter flocks had disbanded. Birds now foraged alone or in small groups, but as soon as the breeding season was over the flocks formed again. At that young age I merely noted the changes in grouping patterns over the seasons. I never gave more than a passing thought to why they should change or why birds gathered in large flocks in the first place. It was not until many years later, when I was a college student, that bird flocks again attracted my attention, but this time as objects of study as well as of enjoyment.

There are many reasons for grouping. Birds may associate passively around patchy environmental resources. For example, individuals gather at watering places in arid areas where surface water is scarce and around safe nesting sites. Reproduction requires some association between males and females and parents and their young: birds must make physical contact to mate, their eggs must be kept warm, and the helplessness of their young demands some parental care. Birds also associate for reasons other than reproduction. Although blackbirds do not cooperate with one another to capture large prey, they do gain information from one another about the location of food. They are attracted to foraging individuals, using them as clues that a particular site is suitable for finding food. More eyes are available to watch for enemies, decreasing the chance that a predator can approach unnoticed. Since birds are very difficult to catch if they initiate their escape while a predator is still at a distance, the advantages of enhanced alertness are great. On the other hand, predators can locate flocks of birds more readily than they can find isolated individuals. Birds in flocks gain protection by avoiding surprise attacks rather than by avoiding detection initially. Also, by grouping together, individuals might be able to defend better some valuable resources, such as a breeding territory or a rich foraging area.

In fact, group living is especially interesting because there are inevitable disadvantages as well as advantages to associating with others. Clustered individuals must share feeding areas, and unless they can cooperatively hunt prey they are likely to reduce food availability for one another. Close association with other individuals also favors transmission of diseases. Combatting human diseases by isolating diseased individuals was practiced long before the causes of diseases and mechanisms of their transmission were well understood. Somehow our ancestors recognized that diseases spread from infected to uninfected individuals and that proximity made spreading more likely. Little is known about the diseases of wild birds, but flocking probably enhances disease transmission in birds just as it does in mammals.

Because of the disadvantages of grouping, there is constant tension among group members over the details of their association, such as how tightly they assemble, the size of the group, and their route of travel. If individuals foraging in a group maintain moderate spacing between themselves, they may still gain the advantages of more eyes watching for predators while reducing the probability that they will interfere with one another by foraging at the same site. The farther apart they are the less mutual interference but also the less likelihood that an associate will spot a nearby predator. In places where predators find it difficult to approach unseen, the advantages of grouping are less than in places where solitary individuals are easily approached undetected. When food is in short supply, the value of reducing interference with feeding is relatively more important than predator alertness, while the reverse is true in times of greater food availability. For these reasons, associations among individuals change with circumstances. There are continual behavioral interactions among group members as each acts to increase the benefits it derives from the association while reducing the disadvantages. These adjustments and interactions provide the best clues to understanding why groups have formed and why they have the features they do.

Nonbreeding Groups among Blackbirds

The flocking for roosting and feeding that I enjoyed

watching as a child is one of the most characteristic features of many temperate zone blackbirds in the nonbreeding season. The size of some blackbird flocks in North America is astounding. A major national blackbird roost survey was conducted in the United States during the winter of 1974-75 by personnel of the U.S. Fish and Wildlife Service with the cooperation of many amateur bird watchers. They counted 723 major roosts with a total of about 438 million blackbirds and 99 million starlings! About three-fourths of this wintering population was found in the relatively humid parts of eastern North America, especially in the lower Mississippi Valley. About 200 million blackbirds and starlings are estimated to winter in Arkansas, Louisiana, Mississippi, and Tennessee alone. There are also substantial wintering populations in the Atlantic Coastal Plain region. In North Carolina eighteen major roosts were estimated to contain about 70 million birds.

In the West, the largest populations winter in the rice-growing regions of California and the Texas coast. Three hundred sixty-five western roosts contained about 139 million birds, about 76 percent of which were in Texas and 10 percent in California.

The sizes of some individual roosts are enormous. In the East, 118 roosts had more than a million birds each; nineteen western roosts were of that size. In Texas, one roost had an estimated 50 million and another had 25 million birds. One roost in Virginia also had 25 million birds, and one in Louisiana had 21 million.

About 190 million (38 percent) of the roosting birds were redwings, followed by 110 million grackles, 99 million starlings (not a blackbird), 91 million brown-headed cowbirds, and 10 million Brewer's blackbirds. Other species found in much smaller numbers were tricolored blackbirds, rusty blackbirds, boat-tailed and great-tailed grackles, yellow-headed blackbirds, and red-eyed cowbirds.

To find food, birds must fly up to one hundred kilometers from these roosts and, even though they must do so only once each day, the time and energy involved may be substantial. How much these wintering birds depress their food supplies is not known, but there is evidence that they sometimes have trouble meeting their energy needs. Daniel Johnson and his associates at Rice University studied wintering brown-headed cowbirds at a roost on this Texas campus. They captured, banded, and measured living birds, and made daily searches of the ground underneath the roost to collect birds that had died the previous night. The number of dead birds they collected during the winter represented only 1.5 percent of the total population, although birds must have died elsewhere as well. Death rates were higher for males than for females and most of the deaths occurred during cold spells. During these periods the

birds had more trouble in finding enough to eat to meet the added demands of maintaining high body temperatures. Interestingly, the males that died were larger than those that survived, suggesting that larger birds had more trouble finding enough to eat than smaller birds.

By an ingenious technique, Robert Selander of the University of Texas compared death rates of great-tailed grackles wintering around Austin. Each evening before going to their roosting sites, the grackles gathered in some large oak trees and sang and called (fig. 5.1). Selander photographed the birds in those trees at regular intervals during the winter. Because males are so much larger than females in this species, he was able to identify the sex of the individuals on large blow-ups of these photos. During the winter males disappeared at about twice the rate that females did. Selander suspected that the major cause of deaths was starvation. The larger males needed to find more food each day than the smaller females, and their very large tails, used extensively in displays in the spring, made it more difficult for them to fly in strong winds. Females were still able to fly to feeding areas during storms that grounded the males.

In winter most blackbirds in North America feed and roost in flocks, but roosting flocks are generally much larger than feeding flocks. Why are such enormous roosts formed? Two factors probably favor flying long distances to roost in such large concentrations. First, blackbirds are vulnerable to predators at night when they cannot see well. Second, they can save a great deal of energy by roosting in places where they are protected from rain and wind and from heat loss by radiation to the cold night sky. Blackbirds commonly roost in marshes where the vegetation is dense and the water makes access difficult for most predators. They also roost in plantations of coniferous trees that provide good protection from wind, rain, and cold. In some areas, blackbirds also roost under the eaves of buildings, but they are less inclined to do so than are starlings, which, as a result, are much greater nuisances to city dwellers.

Flying long distances to favorable roosts, while helping reduce heat loss and deaths from nocturnal predators, raises some problems of its own. Millions of blackbirds quickly deplete food supplies near their roosts, requiring either a change in roost location or longer and longer flights each day to feeding areas. Longer flights are a lesser problem in winter, when only a single round trip must be made each day, than during the breeding season, when hundreds of trips are made each day to a nest with nestlings.

A foraging blackbird usually faces a choice between nearby feeding areas with reduced food supplies or more distant feeding areas with better food supplies.

No studies have been made of how blackbirds decide where to forage, but starlings face the same problems and probably deal with them in much the same way that blackbirds do. William Hamilton and Kenneth Watt studied the distributions of foraging starlings around roosts in the Sacramento Valley of California. They found that birds foraging closer to the roost had lower prey capture rates than did individuals foraging farther away from the roost (fig. 5.2). The differences in feeding rates are just about right to compensate for the differences in flying times so that birds foraging at all distances from a roost have about the same net energy gain. Evidently starlings are not doing a bad job of adjusting their foraging areas to give them the best daily energy balance, but how they do so remains a mystery.

Even though it may be advantageous to roost together, it does not follow that it is better to forage together. Nothing would prevent a bird from leaving its roost and flying somewhere to forage alone. Therefore, group foraging must have at least one of the advantages we mentioned earlier, but which one? How can we tell which of the possible advantages are really the correct ones? We cannot ask birds directly but their behavior may give clues if we are clever enough to look for the pertinent factors.

For example, if the benefit of foraging together is protection from predators, several things should be true. First, birds can spend more of their time foraging and less time looking for predators when in a group than when foraging alone. This can be shown by comparing the behavior of foraging birds in flocks of different sizes. G. V. N. Powell performed such experiments with flocks of starlings and tricolored blackbirds. As expected, members of flocks spent less time looking up and more time foraging than did solitary individuals, and the larger the flock the less each individual looked up (table 5.1). Jamie N. M. Smith, studying flocks of great-tailed grackles in Panama, also found that females paused and looked up less when foraging in more dense flocks than in less dense flocks. Flock density did not, however, similarly affect the surveillance patterns of males. It is not certain why there is this difference between the sexes, but males are always farther apart than females. Being larger than females, they may be at less risk while foraging.

Second, if more eyes offer better protection from predators, then flocks should tend to be larger when birds are foraging in more dangerous areas or when predators are known to be in the vicinity. This possibility has not been tested with blackbirds, but Caraco, Martindale, and Pulliam found that when they released a trained Harris hawk over an area where yellow-eyed juncos were foraging, flocks quickly became larger. In

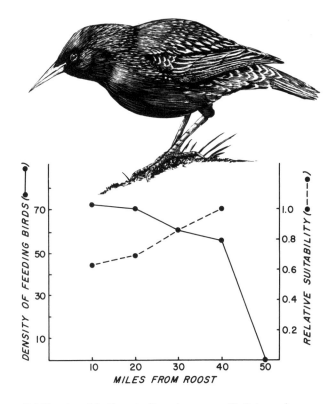

5.2 Density of feeding starlings decreases with distance from the nearest roost, but the relative suitability of the foraging areas increases because the birds deplete food supplies in the fields closer to their roosts.

especially cold winter weather, when birds were hard pressed to find food, flocks became smaller and individuals spaced themselves out more. This suggests that when food is more important, birds are more willing to forgo the advantage of better protection from predators in order to reduce the extent to which they interfere with one another's foraging efficiency.

Third, if flocking is primarily for defense against predators, predators should find it more difficult to attack birds in flocks. Again, there are no pertinent studies of blackbirds, but results from Europe in studies on wood pigeons being attacked by a goshawk illustrate the process. R. E. Kenward found that the percentage of successful attacks decreased as the size of the pigeon flocks increased (fig. 5.3), because birds in larger flocks spotted the goshawk sooner and flew to safety.

Some blackbird flocks are nothing more than loose aggregations at rich food sources, but many flocks are much more tightly structured than that. One of the most striking aspects of many blackbird foraging flocks is the way in which they "roll" across fields (fig. 5.4). This rolling movement is the result of a particular pattern of movement by the flock members. An individual, finding itself at the back end of the flock, flies over its flock-mates to the front and forages in a small area until it again finds itself at the rear of the group.

Table 5.1 **Effects of Flock Size on Surveillance in Starlings and Tricolored Blackbirds (after Powell, 1974)**

| | Flock Size | | | |
	1 Starling	5 Starlings	10 Starlings	1 Starling and 9 Blackbirds
Number of Flocks	6	6	6	7
Percentage of foraging time spent feeding	53	70	88	85
Percentage of foraging time spent looking up	47	30	12	15
Number of look-ups per second	.39	.30	.19	.27

Each individual thus alternates foraging in a relatively small area with short flights to the front of the flock. The result is a continuous flow of birds across the foraging area. The advantage of the system is that each bird has exclusive use of foraging sites that have not yet been visited by other flock members and yet has to make only short flights between feeding locations.

Breeding Groups of Blackbirds

Most blackbirds change their social relations when they are about to breed. At temperate latitudes this occurs in late winter or early spring. In tropical regions it usually occurs when the rainy season begins and the increase in abundance of insects makes conditions favorable for raising offspring. Blackbird breeding social systems are highly varied but fall into a few major categories.

If the breeding unit, whether a pair or larger group, occupies an exclusive area from which other units are expelled, the system is called *territorial*. When the territories of adjacent units directly abut, it is usually easy to observe interactions at the boundaries. Territorial systems range from those in which the birds are regularly spaced over the environment to those in which the territories are clumped in areas where food or nest sites are concentrated. *Grouped territories* are usually relatively small and interactions among the territory holders may be intense, but often those groups are well isolated from other such groups by intervening areas of less suitable habitat.

As clumping becomes more extreme and territories are much smaller, a point is reached when most ornithologists would no longer call the system territorial but would refer to it as *colonial*. There is no unambiguous dividing line between grouped territories and colonies, but the division is a useful one because birds in colonies obtain little of their food within the colonies whereas territorial individuals get most or all of it within the confines of their territories.

A similar but different form of breeding social

organization results when the breeding units occupy *home ranges* that overlap to some degree with the home ranges of adjacent units. Most home ranges have a core area, usually the immediate surroundings of the nest, that is used exclusively by the resident unit, but much of the area used for foraging may be shared with other individuals or groups.

Blackbirds are excellent subjects for the study of breeding-season social organization because their systems include nearly all those that birds employ. Sometimes very closely related species have strikingly different systems, indicating that changes in social organization can and do evolve very rapidly in response to different environmental conditions. Because they are so evolutionarily dynamic, social systems tell us more about current environmental relations than they do about ancestry.

The spatial organization of breeding blackbirds in relation to habitat is summarized in table 5.2. Many species, especially the tropical ones, are yet to be studied during the breeding season, but additional studies are not likely to change the general pattern shown in the table. During the breeding season most blackbirds are territorial. Territories may be highly clumped, as in many of the marsh-nesting species, whose breeding territories do not provide all of the food eaten by the birds and their young. In the dispersed territories, such as those of the forest and grassland species, the territory provides most or all of the food used by the breeding pair or larger group.

Colonial icterids are found in tropical forests (oropendolas and caciques), in a variety of open habitats that may loosely be called savannas, and in marshes. Territorial and colonial species often nest in the same marshes. The highly colonial tricolored blackbird nests in many of the same California marshes with redwings. In marshes of northern Costa Rica I found colonial Nicaraguan grackles nesting together with redwings. In eastern North America common grackles often nest in small colonies in marshes among

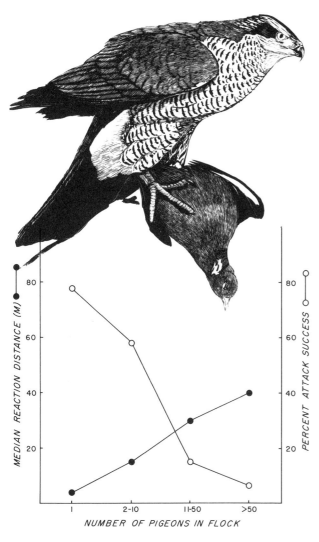

5.3 Goshawks are less successful when they attack wood pigeons in flocks than when they attack solitary pigeons.

5.4 A flock of blackbirds "rolling" across a field, foraging as they go.

redwings. As we saw in Chapter 3, grackles tend to exploit uplands for food more than do redwings. Typically the colonial species use marshes as relatively safe places to build their nests rather than as sources of food.

Among most species of blackbirds, the breeding unit is a single pair that confines its activities within the boundaries of its territory. This is true for orioles, meadowlarks, the melodious blackbird, and the scarlet-headed blackbird. Territorial males of some species, such as redwings and yellowheads, have a number of females nesting on their territories. In still other species, such as the Bolivian blackbird, the breeding unit is a small flock that defends the territory. These interesting aspects of social organization form the focus of Chapter 8. Our immediate concern is the spacing patterns themselves and how they are adapted to different kinds of environments.

The denseness of packing in blackbird colonies is highly varied. The colonies of Brewer's blackbirds are loosely structured. In the Potholes, distances between nests range from several meters up to over one hundred meters. In contrast, a male tricolored blackbird defends a tiny territory (about two-three meters on a side) within which several females may build nests. Colonies of over one hundred thousand nests occupy areas less than a square kilometer.

The Adaptive Significance of Grouping Patterns

The forces that have influenced the evolution of these varied social sytems can be better understood by posing a set of questions. Why defend space? Why not defend all of one's feeding area? Why join together with other individuals of the same species to form colonies? These are not completely distinct questions. Rather, they focus attention upon different aspects of the interrelated decisions that individual blackbirds make.

Why defend space? Like any other activity, defense of space has its costs. Time must be devoted to advertising

Table 5.2 **Association between Grouping Pattern and Breeding Habitat among Icterids**

Habitat	Territories	Grouping Pattern	
		Grouped Territories	Colonies
Forests	Jamaican blackbird Yellow-billed cacique Solitary black cacique		Montezuma oropendola Chestnut-headed oropendola Crested oropendola Yellow-rumped cacique Red-rumped cacique Yellow-winged cacique
Savannas	Melodious blackbird Most orioles		Brewer's blackbird Yellow-shouldered blackbird Common grackle Greater Antillean grackle
Grasslands	Bobolink Eastern meadowlark Western meadowlark Red-breasted blackbird Lesser red-breasted meadowlark Long-tailed meadowlark Brown-headed cowbird		
Deserts	Yellow oriole Bolivian blackbird		
Marshes	Scarlet-headed blackbird	Yellow-headed blackbird Red-winged blackbird Yellow-hooded blackbird Chestnut-capped blackbird	Tricolored blackbird Yellow-winged blackbird Brown-and-yellow marshbird Great-tailed grackle Boat-tailed grackle Nicaraguan grackle
Bogs		Rusty blackbird	

one's presence on the territory, searching for violations of the space, and evicting intruders. Also, these activities draw the attention of predators to the territory holder, increasing the chances that the individual will die as a result of its conspicuousness. For territorial behavior to evolve, these costs must be outweighed by benefits derived from reserving the space for exclusive use. A territory may provide important resources such as food, cover, or nest sites, or it may provide privacy for the performance of activities likely to be disrupted if other individuals are around. It does not pay to defend a food source if there is more than enough food to go around. Among birds space itself is not likely to be in short supply because food is usually exhausted long before there is standing room only. Even if resources are in short supply, however, it does not follow that birds will compete for them by defense of space. For a territory to be profitable, the resources must also be *economically defensible*; that is, a defender must gain

more than it loses. Some resources are indefensible because they are too unpredictable in space and time. For example, the locations of insects flying above forest canopies change with wind direction and speed, cloudiness and temperature. On a particular day a given area may have no insects while on the next day it may be full of them. Such a resource is not economically defensible. There are no swallows that defend their foraging areas. Within a forest, however, insects do not shift around as much, and flycatchers are territorial.

Among blackbirds, food supplies and safe nesting sites are the resources for which defense of space is most often profitable. Starvation of nestlings occurs in all species that have been studied in detail, and predators are a major source of mortality of eggs and young. The importance of food supplies is revealed by correlations between territory sizes and food abundance. More information is available on the territories of red-winged blackbirds than for almost any other bird and, because of its extensive range from the Atlantic to the Pacific and from Alaska to Costa Rica, the redwing can be found in very different types of environments. Also, within a particular region, marshes vary widely in their production of food. There are many opportunities to observe relationships between food supplies and territory sizes in this species.

In the state of Washington the average size of redwing territories on different marshes ranges from 332 to 2,316 square meters. The largest and smallest individual territories fall well outside those limits. These differences, however, are not correlated with measures of lake productivity. At the Turnbull National Wildlife Refuge, territories are smaller on Beaver Pond (average 429 m^2) than on McDowell Lake (average 1,212 m^2) even though hardly any aquatic insects emerge from Beaver Pond while many emerge from McDowell. Mann Lake, which is close to McDowell, has territories of similar size (average 1,207 m^2) even though Mann, like Beaver, has almost no emergence of aquatic insects. The reason is that redwings forage extensively off their territories. In fact, females from Mann Lake often fly to McDowell to forage and females from Beaver fly to nearby Kepple Lake, which is very productive. Redwings breeding in uplands in New York State have territories that average over three times as large as the territories of birds nesting in nearby marshes that are richer in food supplies.

Yellow-headed blackbirds also forage off their territories but less so than redwings. Therefore, they should be more sensitive to lake productivity in determining their territory sizes than are redwings. Nonetheless, the smallest yellowhead territories occur where there are high-quality, undefended foraging areas adjacent to the nesting marsh. In the Imperial Valley of California, Mary Willson found yellowheads nesting at such densities that they could properly be called colonial. They were gathering all of their food in adjacent croplands. In the Cache Valley, Utah, I observed yellowheads nesting at very high densities in small cattail islands in a lake formed by a dam on the river. They were foraging on nearby irrigated cropland, especially when alfalfa was being harvested and the cover was literally being cut out from under the insects. Even in the Potholes, where the uplands are generally dry and unproductive, the densest populations of nesting yellowheads occur where there are irrigated pastures next to the marshes. In May 1981 I counted 125 active nests of yellowheads in a patch of cattails and bulrushes only about 1,000 square meters in area. In several places I could touch three or four nests without shifting position. Birds nesting in that marsh foraged extensively in irrigated pastures.

There are, however, marshes at the Potholes with no pastures nearby. James Wittenberger made an intensive study of food production on territories of yellowheads on several different lakes in 1978 and 1979. He found that territories were smaller and the density of nesting females was higher on the more productive lakes. However, he did not find any correlation between food production on a territory and the number of females settling on that territory. Females used productivity to decide which lake to settle on but not to decide which territory. When picking territories they were more concerned about protection from predators for their nests.

Male bobolinks defend large territories within which a large proportion of the food for the adults and nestlings is gathered. In eastern Oregon James Wittenberger measured both the sizes of bobolink territories and the production of food on them. He found a rather poor correlation between food abundance and territory sizes because the males had to select their territories long before the food upon which breeding depended was available or could be estimated accurately. Females, however, selected their nesting sites much later and were able to use more information than was available to the males. Their nesting densities showed a better correlation with food abundance than did the sizes of the male territories.

Why not defend the whole space? As illustrated by redwings and yellowheads, many icterids defend territories that contain some but not all of the resources required for breeding. If an individual exerts the effort to defend some of its feeding area, why not all of it? Part of the answer is that costs of defense increase with the size of the area being defended. As long as benefits rise faster with area than do costs, a larger territory will be advantageous, but the value of holding more space gradually decreases (fig. 5.5). Food sources farther from the nest are less valuable than food sources closer to the nest because more time and energy must be expended in flying to and from them. Also, patrolling costs increase with the size of area defended. The larger the area the more difficult it becomes to detect intrusions. As incursions into the territory increase, fewer resources are reserved for exclusive use of the owner. Therefore, at some territory size costs start to rise faster than benefits. Defense of a larger territory actually yields lower net benefits to the defender than defense of this optimal size. Whether or not the best size of territory includes all of the needed resources depends on resource density. It also depends on whether or not there are undefended areas around the territories where no individuals can profitably exert control. Uplands and edges of lakes with emergent vegetation are such undefended areas and breeding blackbirds use them extensively.

Why form colonies? As we discussed in Chapter 3,

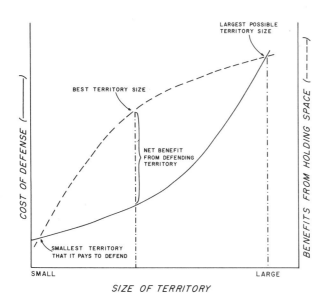

5.5 *A cost-benefit model of territory size. The costs of defending a space increase with its size more rapidly than the benefits. As a result, there is an optimal size where the net benefit from defense of the space is greatest.*

food resources that are patchily distributed in space and time are exploited most efficiently from a central location. All birds seeking the best places for their nests choose the same site and a colony forms. If there are only a few safe nest sites, the birds gain no particular benefits from being close to one another—they gain by being in that particular spot whether or not anybody else is there. However, individuals may benefit from one another either because of group defense of nests against predators or because the individuals use one another as clues for finding places with good food supplies.

Colonially nesting blackbirds do join together to attack predators near their nests, but individuals attack most strongly near their nests and progressively more weakly farther away. A harrier (marsh hawk) flying through a colony of Brewer's blackbirds is mobbed progressively by a different set of individuals as it passes over. A stuffed horned owl placed within a colony is mobbed strongly by individuals with nests close to it but is largely ignored by individuals whose nests are farther away (fig. 5.6). In all of our studies we have not found any evidence that group mobbing achieves effects different from those resulting from mobbing by a single pair. In fact, nearby individuals take advantage of the opportunity presented by preoccupation of the nest owners with the predator to court females or harvest resources. Group welfare does not appear to be on the agendas of individuals.

Fishermen are famous for using one another as sources of information about where to fish. On Puget

Sound a bent fishing rod, combined with vigorous activity on the part of its holder, is a powerful magnet swiftly attracting many other boats. Blackbirds also use one another as clues for foraging. The individuals are not deliberately signaling to one another any more than a fisherman is signaling with his bent pole, but their actions provide information to others. The food supplies used by blackbirds are harvested efficiently by solitary individuals, and there is no foraging benefit to attracting companions. Nevertheless, a bird returning to its nest with a large load of food in its bill provides some information to other birds who happen to be in the vicinity at that time. It is a relatively simple matter to fly off in the direction from which the loaded individual arrived, keeping a lookout for good sites along the way. The closer the nests are to one another, the easier it is to observe the neighbors' activity.

During his studies of Brewer's blackbirds in the Potholes, Henry Horn observed this behavior many times. Individuals who had not been particularly successful in their foraging trips sometimes waited at the colony observing the arrivals and load sizes of incoming individuals. Evidence that this really pays off is that young in nests in the centers of colonies, where adults have more neighbors to watch, grew faster than nestlings in nests at the periphery of colonies.

The rapid changes in directions from which individuals arrive and depart at icterid colonies also suggest that the birds observe the activities of fellow colony members. In the large colonies of tricolored blackbirds in California, the directions in which birds leave to forage change many times during a day, and at any one time most of the birds are usually departing in the same direction. Some of these shifts of direction are responses to changes in the surrounding foraging areas such as flooding of a pasture or cutting of some crop. At a colony of yellow-rumped caciques in Peru, I also observed frequent shifts in foraging directions (see fig. 3.10). These shifts in direction may have been caused by discovery of some new fruiting tree, or the swarming of some species of insect, but I was unable to follow the birds to the foraging areas to determine what they were finding; nor could I identify food items in their bills when they arrived at the nesting colony.

How big should colonies be? Because it is good to have some close neighbors does not mean that it is good to have a large number of them, but nesting birds usually cannot control the behavior of newly arriving individuals. If newcomers really want to settle, it will be hard to prevent them from doing so. Established residents have eggs or nestlings that need care, and a persistent arrival can easily raise the costs of exclusion above the price a resident can afford to pay. Nonetheless, colonies do not increase indefinitely in size, suggesting that newcomers are making decisions

5.6 *Brewer's blackbirds mobbing a great horned owl sitting close to their nests.*

about adding themselves to an existing colony or starting a new one of their own.

Finding out how they make these decisions is very difficult, but clues can be gleaned from the birds' behavior. The sizes of colonies in relation to environmental conditions is informative. Among tricolored blackbirds, for example, colonies are larger in the very productive rice-growing areas of the Sacramento Valley than they are in the less food-rich foothills (table 5.3). This makes sense because in less productive environments birds must fly farther to find food than birds in a similar-sized colony in a more productive environment. The behavior of tricolored blackbirds when they are establishing a colony suggests that they are assessing the resources of the surrounding region. They regularly fly out from a potential nesting area to feeding sites and back again. These trips are much more frequent than would be needed simply to feed. Tricolors may also abandon nesting attempts after eggs have been laid, probably on the basis of updated estimates about the environment.

Table 5.3. **Sizes of Breeding Colonies of Tricolored Blackbirds in California**

Breeding Area	Number of Colonies with		
	less than 1,000 nests	1,000–10,000 nests	more than 10,000 nests
Foothills	7	2	1
Valley cropland (no rice)	3	2	3
Rice country	0	3	7

The size of blackbird colonies is rarely determined by the number of suitable nesting sites. In colonies of oropendolas, there are invariably more branches on which nests could be slung. Colonies of Brewer's blackbirds in the Potholes seldom fill all of the area of suitable sagebrush. Tricolored blackbirds rarely fully occupy an alfalfa field or patch of marsh. The limits of colony size are set by the responses of birds to resources and varying quality of nest sites, but we have much to learn about how they make their assessments.

6.1 A female yellow-rumped cacique weaves material into her
nearly completed nest while the male, who never assists her,
sings and advertises nearby.

6. Nests and Nest Sites

The nests of birds have long attracted the attention, admiration, and even appetites of people. The ancients, limited by their level of scientific understanding, tended to interpret the nest-building behavior of birds anthropomorphically. In the regular comings and goings of birds with nest material they saw a sign that neither their nor our behavior was random or would go unrewarded. The scientific interest in birds' nests is more recent but, luckily, we have not lost our sense of awe at the achievements of birds as builders.

John Muir, in his engaging book *The Mountains of California*, writes lovingly of the nest of the water ouzel.

The Ouzel's nest is one of the most extraordinary pieces of bird architecture I ever saw, odd and novel in design, perfectly fresh and beautiful, and in every way worthy of the genius of the little builder. It is about a foot in diameter, round and bossy in outline, with a neatly arched opening near the bottom, somewhat like an old-fashioned brick oven or Hottentot's hut. It is built almost exclusively of green and yellow mosses, chiefly the beautiful fronded hypnum that covers the rocks and old driftlogs in the vicinity of waterfalls. They are deftly interwoven, and felted together into a charming little hut, and so situated that many of the outer mosses continue to flourish as if they had not been plucked.

Accounts of the details of construction, while less poetic than John Muir's prose, nonetheless give us a vivid impression of the intricacy of the process. Francis Herrick described nest building by a Baltimore oriole as follows:

Clinging to the principal twig, hanging often with head down, and holding the thread, the bird makes a number of rapid thrust-and-draw movements with her mandibles. With the first thrust a fiber is pushed through the tangle which soon arises and forms a growing mass, and with the next either that or some other fiber is drawn loosely back. . . . While these shuttle movements are, first and last, very similar and almost equally rapid at all times, the number made at each visit tends to increase with the growing complexity of the product. At least one hundred shuttle movements were sometimes made at a single visit, but these were often so rapid that it was impossible to count them, and many of them must have been abortive. . . . In all this admirable work there was certainly no deliberate tying of knots, yet, as the sequel will show, knots were in reality being made in plenty at every visit.

Building a nest is certainly one of the most complex behavioral acts performed by any animal. Hundreds or thousands of pieces of material must be arranged in such a way as to give a receptacle of the proper size,

and they must be interwoven sufficiently to withstand the pressures that will be exerted on them by the weather, the incubating bird, and the activities of the nestlings (fig. 6.1). As most small birds do not re-use their nests, the construction need only withstand those pressures for a month or two, but shabby construction can cause loss of eggs or young. The hundreds of trips needed to build a nest may be accomplished in as little as three to four days. This high level of activity makes the builder conspicuous to friend and foe alike. Predators take 30 to 60 percent of the eggs and young of most small birds. Therefore, selection of sites that offer protection from predators is an important consideration.

Nests may be protected from predators by being hidden in dense cover and by matching the materials surrounding them. The smaller the nest the more easily it can be concealed. There are few places to hide large nests. Most large birds depend on the inaccessibility of their nests or the constant presence of one of the adults to ward off predators. Smaller birds can also choose inaccessible sites, but they can defend against fewer predators than can large birds. Nest defense is a conspicuous activity among some small birds, but it is difficult to judge how effective it is.

A nest also influences the temperatures to which the eggs and young are exposed. A thick, well-insulated nest provides protection against wind, cold, and solar heat, supplementing the efforts of the incubating adult. In such a nest, temperatures are likely to be above those of the surroundings during the night but below them during the day, providing the contents with a more equable environment. Nests built in cavities and holes provide the most constant temperatures.

As early as 1842 John James Audubon noted that Baltimore oriole nests in Louisiana were more loosely woven than those of orioles nesting farther north. He speculated that the open weave allowed the wind to circulate through the nest more readily. This would be effective only if the Louisiana orioles placed their nests in the shade, because a flimsy nest heats up more in the sun. Recent studies of orioles in the Great Plains have shown that nests are not placed to avoid exposure to the sun but rather are concentrated on the west side of trees where they are exposed to the strong prairie winds. The insulative qualities of nests have been tested by putting them in a plywood chamber in the laboratory with strong heat lamps. Temperatures inside

6.2 *Arid valleys on the eastern slope of the Andes Mountains in Bolivia are the home of the rare Bolivian blackbird. Members of the flock holding this area fed mostly in the shrubs on the sides of the valleys.*

the nests were recorded with the fine tip of a thermometer positioned about two centimeters above the bottom. Oriole nests from the southern plains, which are exposed to very hot temperatures during the breeding season, have better insulative qualities than those from the northern plains. They are thicker and bulkier, keeping heat absorbed from the sun on the surface of the nest where it can be dissipated by the wind.

The catch is that better insulated nests are larger and easier for predators to locate. By constructing sturdier nests birds reduce one problem but aggravate another one. It may not matter to orioles whose nests are located on the very tips of slender branches where they are relatively inaccessible even if predators do know where they are. Also, if a nest is already conspicuous, adding more material may not increase the chance that it will be discovered. In fact, the added material could both increase insulation and make it more difficult for predators to enter, especially if the nest is a closed one that predators cannot rob without ripping apart .

The Nests of Blackbirds

Most species of blackbirds build open, cup-shaped nests, either on the ground or in a crotch in a tree or shrub. The ground nests of meadowlarks are domed so that their contents cannot be seen from directly above. The nests of other species, particularly the marsh-nesting ones, are not placed in a crotch but are, instead, suspended among several vertical stalks. When redwings build their nests in trees and bushes, they hang them just as they do in cattails. The most complex nests in the family, built by orioles, caciques, and oropendolas, are elaborate, pendant nests which are normally suspended from the tips of tree branches. Entrance to these nests is through a hole near the top. Finally, a few species place their open cup-shaped nests in cavities in trees or holes in cliffs.

Hole nesting is very rare among blackbirds. Redwings have once been reported nesting in a bird box, and Brewer's blackbirds occasionally nest on rock ledges. In Puerto Rico, the yellow-shouldered blackbird

6.3 Nesting site of the Bolivian blackbird (indicated by a white arrow) on a cliff beside a stream channel now dry but full of water in the rainy season.

usually builds an open nest, but some nests are built in cavities in dead snags. The only icterid known to breed exclusively in holes is the Bolivian blackbird, an inhabitant of rugged arid valleys on the eastern flanks of the Bolivian Andes. In that environment vegetation is sparse (fig. 6.2), and large nests in the small trees and shrubs would be very conspicuous to predators. It is therefore not surprising that this species has evolved to conceal its nests in crevices and holes in these cliffs (fig. 6.3). The birds must fly further for food for nestlings than they would if they nested in trees, but safety probably more than compensates for this extra effort. The Bolivian blackbird is also unusual because it breeds in small flocks and more than two birds feed nestlings in each nest.

Some species of blackbirds are very predictable in both the location and the construction of their nests while other species all choose the same types of nest site. Why does this variability exist? Among the better-known species, those that regularly forage in a broader range of sites also nest in a wider variety of sites than

do those species with more specialized foraging modes.

For example, yellow-headed blackbirds virtually always nest over water. Most of their nests are built in emergent herbaceous plants, but occasionally they use shrubs and small trees growing in the water. Yellowheads, as we have seen, breed only on high productivity lakes and marshes. They raise their young mostly on aquatic insect prey and they seldom forage in trees.

Red-winged blackbirds, however, build their nests in emergent aquatic vegetation, trees and shrubs growing over dry land, and even on the ground. The way the nest is constructed changes with the site (fig. 6.4). Redwings are very catholic in their foraging. Not only do they forage in all of the places where yellowheads do, but they also obtain food for their young in trees and shrubs. They also breed on marshes of low productivity and in fields and croplands where all of their foraging is done over dry land. Brewer's blackbirds build their nests in many different places, the preferred sites changing with locality. In the

6.4 Nests of the Baltimore oriole (above) and red-winged blackbird (below), showing how the details of construction change with the nest site even though the general form of the nest does not.

Potholes most nests are constructed in sagebrush, but many are on the ground. A few are built in emergent vegetation over water, and very rarely the birds nest on ledges on the basaltic cliffs. In the Cariboo Parklands of British Columbia nearly all Brewer's nests are built on the ground, while in parts of California most nests are built in trees, particularly pines. Brewer's blackbirds exploit aquatic insects heavily, but they also forage in a variety of grassy and sedgy places and are inclined to build their nests in any relatively safe site affording good access to feeding areas. They would probably nest more in marshes than they do if they were not excluded by redwings and yellowheads. In fact, Robert Furrer found that Brewer's readily used sagebrush he anchored in the water at the Potholes, in places where the shoreline was undefended by other blackbirds, making access easier for the Brewer's.

Orchard orioles usually build their shallow, hanging nests in trees and large shrubs, but in the delta of the Mississippi River they commonly breed in marshes, where, like redwings, they build their nests in reeds. The foraging sites of the adults in this environment are not known.

There are places where blackbirds are unable to breed because of the lack of suitable nest sites. In many areas of western North America, there are lakes so alkaline that cattails and bulrushes cannot grow in them. These lakes are often highly productive, in fact more productive than many of the lakes on which blackbirds do nest. But they do not offer good nest sites, so blackbirds use them only for foraging.

In eastern North America common grackles tend to nest in coniferous trees planted in parks, gardens, and residential areas of cities and towns. Their rather bulky nests are conspicuous in deciduous trees, especially since they nest early in the spring before the trees have come into leaf. Grackles are much more common in many areas of the central United States than they would be if people had not planted good nesting trees for them. When marshes are present, grackles regularly build nests there, but they tend to gather their food in the uplands, using the marshes primarily as safe nest sites.

How Do Blackbirds Decide Where to Put Their Nests?

One can imagine several ways in which birds could decide where to put their nests. One would be a rigid inherited behavior that specified the type of nest site that an individual used. Such a system is unlikely because nest sites are variable in type and quality, and it is usually better to accept a less than optimal site rather than not nest at all, provided the site offers some chance of success.

Alternatively, birds may learn which sites to use by observing where they were raised. Before leaving its nest the individual could learn the characteristics of that site and seek similar sites when it is ready to breed. Such a system is not unreasonable because successful fledging from a nest constitutes *prima facie* evidence that the site was good. This possibility was tested (but found to be incorrect) by Robert Furrer with Brewer's blackbirds in the Potholes. He banded hundreds of nestlings from nests in sagebrush, on the ground, and in sedges over water. In subsequent breeding seasons he searched for banded females to see where they built their nests. The type of nest in which a female was raised did not influence where she built her own. Females raised in ground nests were just as likely to build nests in sagebrush as were individuals reared in sagebrush, and *vice versa*.

Flexibility in choice of sites should be advantageous for a number of reasons. First, a nestling does not know where it will eventually settle to breed. Therefore, it cannot know what types of sites will be available to it. In addition, availability of sites changes from year to year even in the same area. Fires or other disturbances may alter vegetation. The vegetation may change without any outside influences as grasses grow taller and shrubs give way to trees. Finally, there may be such competition for nest sites that when an individual is seeking a site the best ones may already be taken.

Methods of nest construction are more constant than the places in which they are built. A redwing nest is a redwing nest and easily recognizable, whether it is built in cattails, on the ground, in a tree, or in a shrub. When I went to Costa Rica to study them in tropical marshes I had no difficulty recognizing the first redwing nest I found. It would not have looked out of place in Washington, Wisconsin, or Alberta. Yellowhead nests are always built of cattail strips and lined with still narrower strips of cattails. Orchard oriole nests are always constructed of fresh green grasses that dry to give the nest a very characteristic light brown color.

Such conservatism in nest structure is probably owing to the fact that the construction materials used by most small birds are abundant and widespread in all environments in which they nest. If these typical materials have some special value, it is not difficult to find them. There is seldom a need to use alternative materials that may not serve needs of concealment and thermoregulation as well. Commonly available materials are not necessarily easily harvested. Alexander Skutch describes watching a female Montezuma oropendola extract strips from a large banana leaf. Standing on the midrib, she pecked the underside of the rib with the sharp point of her bill until the fibers became separated. Then she grasped one of the exposed sturdy fibers with her bill, and ripped out a piece up to two feet in length. The entire process took many minutes.

Orioles and oropendolas build the bulk of their nests by hanging upside down on the sides of the pouch and doing all of the weaving with their bills. Only after the cup is formed do the birds apparently use combined movements of bill and feet to manipulate materials. Oropendolas hold leaves in their feet and shred them with their bills to produce the fine materials that line their nests.

For large birds suitable nest materials may not always be readily available and more striking substitutions of materials do occur. In some of the desert regions of the American Southwest, common ravens have built their nests out of barbed wire lined with horsehair. In eastern Washington they sometimes use sheep rib bones and line the nest with shredded sagebrush bark. These unusual nests are built in areas where the sticks which ravens usually use are difficult or impossible to find but where there are food supplies to support breeding.

Who Builds the Nest?

In most orders and families of birds the nests are built by both sexes, but among song birds patterns are quite variable. The nests of blackbirds are built entirely by the female in most species. Male and female melodious blackbirds share in building their nest. In a few species, such as the brown-and-yellow marshbird of Argentina, males occasionally pick up nest materials and may even take them to the nests, but usually they drop the pieces after holding them for a while as if they do not know what to do with them.

In only one blackbird is the male known to construct the nest in its entirety. The male yellow-hooded blackbird of northern South America usually concentrates on one nest at a time (fig. 6.5), but if he does not succeed in attracting a female to that nest he may abandon it and start on a new one. A special fluttering flight is used to lead females to these nests, which are concealed in the rank vegetation of marshes and wet meadows.

Nest-site demonstration displays are also performed by males in species in which the female alone builds the nest. Redwing males frequently fly down into clumps of cattails and utter very harsh, guttural sounds. Females follow males into the cattails and may build in sites pointed out by the males. Tricolored blackbird males, who defend tiny territories in their dense colonies, spend a great deal of time in a wings-up, nest-site demonstration display very similar to that of redwings. These displays suggest that the presence of good nest sites is an important component of a territory and that the females' perceptions of those sites

may enhance a male's success in attracting a mate.

The preponderance of nest building by female icterids is somewhat puzzling because in the African weaverbirds, a group that is in many ways very similar ecologically to blackbirds and has many parallel features among their social systems (table 6.1), nest building is done almost entirely by the males in most species. Why should they be so different? To approach this question we need to consider the costs and benefits of nest building to an individual.

One obvious cost is the time and energy expended. The time devoted to gathering material as well as to constructing the nest is substantial and represents time taken away from other activities of value, among them feeding. In addition, flights to and from the nest are about twenty times as expensive in energy as sitting or walking. The builder needs more food per day but has less time in which to gather it. At the same time that a female is building a nest she should also be accumulating extra energy for her eggs, which are normally laid as soon as the nest is completed. Nest building and foraging compete strongly for her time. For males, other types of activities, such as defense of territory, courting additional females, and watching for predators, compete most strongly for time.

The benefits from building a nest include its potential use as a roosting site outside the breeding season, although this has been observed only once among blackbirds. Alexander Skutch observed troupials roosting in a nest of the rufous-fronted thornbird at least a month before they began to breed in it and for a long time after their young had fledged.

To a male, the most important benefit of nest building is the increased chance of attracting more and better-quality females. A female could be attracted to a male with a nest because she can save time and energy by using it. Second, the presence of the nest may be a good indication of the vigor and quality of the male. By building a nest a male demonstrates that he is such a good forager and effective competitor that he can take time out to construct a nest and still be in good enough condition to hold his space and court her energetically. This value of a nest should be especially high if the male's territory does not provide any food and the female is making her decisions primarily on the quality of the male.

A nest may also advertise the presence of its builder even when he is absent. It may serve to repel invading males and cause prospecting females to delay longer in the area. These advantages seem important enough that we would expect males in most species of birds to build nests, but this is not the case. It is much more common, especially among smaller birds, for females to build the nests unaided by their mates. Evidently the factors that have molded nest-building behavior have

6.5 *A male yellow-hooded blackbird works at completing a*
nest while a prospecting female observes his progress.

Table 6.1. Comparison of Social Organization among Marsh-Nesting Blackbirds and Savanna-Nesting Weaverbirds in Africa

Activity	Blackbirds	Weaverbirds
Breeding aggregations	Highly clumped	Highly clumped
Territories	Small to medium	Limited to area around nest
Mating systems	Mostly polygynous	Mostly polygynous
Nest construction	Female only except in one species	Mostly by male
Incubation	Female only	Both sexes
Feeding of young	Both sexes but more by female	Both sexes
Method of food delivery	Carried in bill	Often regurgitated

not included being helpful to one's spouse! Individual benefits seem to have been the driving forces. But why should they lead to such different patterns in blackbirds and weaverbirds?

We do not know, but the following argument is plausible. In tropical savannas where most weaverbirds breed, food availability is lowest at the end of the dry season and beginning of the rainy season. During the dry season the seeds upon which the birds feed are gradually depleted and those that do remain suddenly convert themselves to nonfood by germinating when the first rains fall. Food supplies subsequently increase during the rainy season as populations of insects build up and grasses mature and develop a new crop of seeds. Young weaverbirds are fed on insects and on new seeds when they are still in the milky stage. Colonies of weaverbirds are located in scattered savanna trees, and in most species males do not defend territories that provide any food. Rather, they defend a small portion of a tree branch that provides only nest sites. Since the males are not defending a large territory, there is little conflict between nest building and holding space. If females are under strong pressures to accumulate energy reserves in order to be able to breed as soon as possible after the first rains fall, then the value of a nest to them should be very great. Also, the presence of a nest should indicate a great deal about the condition of the male.

Males of most blackbird species defend territories that contain at least some of the food supplies for breeding. Defense of these areas is more costly of time and energy than is defense of a tree branch. Moreover, by defending a good area a male both provides resources for females and informs them about his success in holding high-quality resources. Therefore, the value of the nest in helping females discriminate among males is probably less for blackbirds than it is for weaverbirds.

Also, in spring in temperate latitudes, there is a gradual increase in food supplies rather than the sudden changes that accompany the transition from dry to wet seasons in the semiarid tropics. During this period it may be easier for females both to build up their energy reserves and to begin to construct nests than it is in the tropics where quick initiation of nests is so important. This line of argument is supported by the fact that the only blackbird species in which males are known to build the nest—the yellow-hooded blackbird—breeds in tropical regions with marked wet and dry seasons. A good test of this idea would be to determine who builds the nest in some of the other tropical members of the genus *Agelaius* that are still virtually unknown. A little piece of evidence suggests that the study would be worthwhile. On December 15, 1970, southeast of Buenos Aires, Argentina, I found a pair of chestnut-capped blackbirds in a small roadside marsh. The female was incubating two eggs while the male was building a second nest about twenty-five feet away. Within about fifteen minutes he made two trips with material to the one-fourth completed nest.

A comparative approach, in which we examine the behavior of many species in different types of environments, is often the only way in which hypotheses such as the one we have just discussed can be tested. If all members of a species behave in the same way, as is usually the case, we cannot use comparisons within species to enlighten our understanding of the behavior. If, however, some males do and some males do not, or if the proportion that do so varies from one habitat type to another, then we could compare proportions in different environments or examine the success of individuals with different behaviors within an environment to determine if the patterns support or reject the hypothesis.

How Do Birds Learn to Build Their Nests?

If there were ever an act where it would be helpful to have a good teacher, nest building would be it. However, opportunities to watch another individual building are rather limited. By the time a young bird is old enough to observe a nest being built, most of the building for that breeding season is finished. Young birds could examine the nest in which they were raised, but the structure itself does not reveal very much about the processes by which it was assembled.

Alternatively they could wait until the following breeding season to observe other, more experienced, individuals before starting their own building. In many temperate-zone birds, young females do nest later than older ones so that they could make such observations, but there is no evidence that they actually do so. Captive birds that have never seen a nest being built usually build a reasonable facsimile of one the first time they try. Observational learning plays, at most, a very minor role in the development of avian nest-building skills.

There is some evidence, however, that birds building for the first time may do a poorer job of it than experienced individuals. A female yellow-headed blackbird was observed to build four nests before she attained one that was usable. It was suspected, but not definitely known, that she was building nests for the first time.

For the elaborate hanging nests of orioles and oropendolas, construction of the loop or ring of fibers from which the nest hangs seems to be the most difficult part. Frank Chapman noticed that some female chestnut-mantled oropendolas built these rings very quickly while others spent days on this part, trying several times before they constructed a suitable one. Alexander Skutch observed the same thing with Montezuma oropendolas. Both of these observers thought, but were not sure, that the individuals having difficulty were breeding for the first time.

The success of inexperienced birds in building nests demonstrates that this complex behavior can be coded into organisms so that they develop the skill without the prior opportunity to observe construction. If other complex behaviors are learned, there must be specific advantages in having them learned. Nest building evolved to be mostly preprogrammed behavior because there are few opportunities for most individuals to learn by watching others, and because the short life spans of most birds means that, on the average, an individual constructs very few nests before it dies. If the individual does not do a good job the first time, it may lose half or more of its expected reproductive output. Learning is a luxury that cannot be afforded under those circumstances. Yet there is flexibility, and performance does improve with practice. This is necessary if nests are to be adapted to the particular sites in which they must be built.

7.1 The cowbirds. Clockwise from top left: shiny cowbird,
screaming cowbird, bay-winged cowbird, bronzed cowbird,
and brown-headed cowbird. The giant cowbird is in the center.

7. Parasitism

Many small birds are tricked each year into raising the young of other species that have hatched from eggs that the brood parasites have laid in the nests of their hosts. Brood parasitism has evolved independently a number of times in the avian world, but only about 1 percent of all species are brood parasites. Included in this interesting group of birds are about half of the 140 species of cuckoos, two genera of African finches, the South American black-headed duck, a small group of rather drab African birds, the honeyguides, and 5 of the 6 species of cowbirds (fig. 7.1).

Cowbirds have received a great deal of attention as ornithologists have attempted to understand why such unusual breeding behavior evolved, what maintains it, and why the host species do not evolve effective defenses against this foreign intrusion.

Although most attention has been directed toward *interspecific* brood parasitism, *intraspecific* brood parasitism, laying eggs in nests of other individuals of the *same* species, is much more widespread. The difficulty is to detect it. Direct observation is the best method, but this is very difficult because it requires that the female that built the nest be distinguishable from the intruder. Indirect evidence is a bit easier to collect. It includes observing two or more eggs being laid in a nest in one day (most small birds lay one egg each day until the clutch is complete), appearance of new eggs after a clutch is completed, the existence of abnormally large clutches, and eggs that hatch noticeably later than others in the nest.

Such indirect evidence indicates at least occasional *intraspecific* brood parasitism in 54 species of birds, 32 of which are ducks and geese. Only 6 species of song birds are among them, none of which is a member of a family in which *interspecific* brood parasitism is known to occur. Virtually no evidence exists concerning which females in a population deposit their eggs in the nests of other individuals. Do a few females do so regularly or do many females do so rarely? We do not know.

The difficulty of detecting intraspecific brood parasitism is illustrated by the red-winged blackbird, probably the most intensively studied North American bird. Until very recently there was no evidence that females of this species ever deposited eggs in nests of other individuals. In the Potholes, however, we have found a limited but regular number of nests in which two eggs appeared in a single day, in which an egg appeared more than one day after the clutch was complete, or in which single eggs were deposited after the nest had been abandoned. Only an intensive study, during which nests were visited almost daily, revealed this pattern.

In addition to brood parasitism, other forms of parasitism also exist. The simplest—taking over nests of other species and then using them to rear one's own young—is the form practiced by the bay-winged cowbird of southern South America. Baywings take over nests, often active ones, of many other species, but they always incubate their own eggs and feed their own young. More than one pair of baywings may be involved with a single nest, but as yet we do not know who they are, whether or not they are related to one another, or whether more than one female actually lays eggs in a given nest.

The baywing is the only icterid that breeds exclusively in nests of other species while still raising its own young, but some other members of the family do so occasionally or regularly. The troupial, a bright oriole of South America, often uses abandoned nests of other species. In Ecuador and Bolivia it commonly uses nests of another icterid, the colonial yellow-rumped cacique. In fact, David Pearson, who discovered this behavior, never observed troupials constructing their own nests in either of these two countries. In Brazil, the troupial has been observed to use the nest of a large ovenbird (*Pseudoseisura cristata*), which is domed and has a tubular side entrance. Using nests of other species may be the usual nesting behavior of troupials because nobody has found a true nest of the southern race of this species. It has also been reported that individuals of the northern race usually take over and repair nests of other species. In the arid coastal zone of northern Venezuela troupials sometimes use the nests of yellow orioles and pale-breasted spinetails, but they also build their own nests. So far as is known, they always build their own nests on the islands of Bonaire and Curaçao off the Venezuelan coast.

Such behavior occurs rarely among other species as well. During the spring of 1964 in the Potholes, I found an active redwing nest in a previous year's Bullock's oriole nest in a small willow tree. The nest was relined by the female redwing, and she successfully fledged three young from it.

The other five species of cowbirds are all brood parasites (table 7.1). Two of them, the North American brown-headed cowbird and the South American shiny

Table 7.1 **Known Hosts of the Parasitic Cowbirds**

Species	Number of Host Species	Comments
Brown-headed cowbird	216 recorded	Expanding its range and number of regularly used hosts
Shiny cowbird	176 recorded	Often uses dump nests; expanding its range and number of host species
Red-eyed cowbird	71 recorded	Orioles are the chief hosts; often uses dump nests; half of these host species also used by brown-headed cowbirds
Screaming cowbird	1	Parasitizes only the bay-winged cowbird
Giant cowbird	7	Parasitizes almost exclusively caciques and oropendolas

cowbird, parasitize a very large number of hosts, probably more than two hundred species each, while the other three are more restricted in their use of hosts. The Central American red-eyed cowbird is known to parasitize seventy-one species, but its most important hosts are orioles. The giant cowbird parasitizes caciques and oropendolas almost exclusively, while the screaming cowbird is known to parasitize only the bay-winged cowbird. Thus, three of the five parasitic cowbirds parasitize primarily or exclusively other members of the blackbird family.

The Effects of Brood Parasites on Their Hosts

The best way to judge the effect of a brood parasite is to compare the success of parasitized and unparasitized nests in the same area. With one exception, studies have demonstrated that brood parasites decrease the nesting success of their hosts (table 7.2). Females of some parasites remove one host egg when they deposit their own egg. Young cuckoos often evict all of their nest mates soon after they hatch. Even in such species as cowbirds where the nestlings do not normally molest their nest mates, the large cowbird, by getting its head up higher, garners so much of the food being delivered that its nest mates starve or leave the nest in poor condition. In addition, some parasites destroy eggs that are well along in incubation, presumably to induce the hosts to lay again soon and provide another opportunity for the parasite.

Cowbird parasitism has such an adverse effect on the nesting success of the rare Kirtland's warbler that strenuous attempts are being made to control cowbird populations in the limited breeding range of that warbler species in central Michigan. A similar situation occurs with the yellow-shouldered blackbird of Puerto Rico, which is very heavily parasitized by the recently arrived shiny cowbird. Today, the only yellow-shoulder nests to fledge many young are found on small coastal mangrove islands where cowbirds are absent.

The one exception is extremely interesting. Neal Smith has found that being parasitized can enhance the survival of young caciques and oropendolas in Panamá because the young giant cowbirds remove botfly larvae from their nest mates. In nests with no cowbirds, botflies can cause high mortality among young caciques and oropendolas, unless wasp nests are present in the same tree with the colony. Wasps are very effective killers of botflies, and in such colonies unparasitized nests do better than parasitized ones.

Even more interestingly, Smith discovered that there are two behavioral types of oropendolas and caciques. One type habitually nests in trees with wasps and is hostile toward giant cowbirds. The other habitually nests in trees without wasps and does not attempt to evict cowbirds. There also appear to be two behavioral types of cowbirds. Those that visit trees with wasps are furtive and attempt to sneak into unoccupied cacique and oropendola nests to lay their eggs. Those that frequent colonies without wasps are bold and conspicuous and commonly sit in the colony tree observing the activities of their hosts (fig. 7.2).

Even though cowbirds reduce the success of nests in which they deposit their eggs, they do not necessarily depress overall populations of their hosts, as they do

Table 7.2 **Effect of Cowbirds on the Success of their Hosts**

Species	Host	Percent of Eggs Producing Flying Young	
		Nonparasitized	Parasitized
Brown-headed cowbird	Eastern phoebe	46	9
	Acadian flycatcher	61	54
	Red-eyed vireo	81	22
	Kirtland's warbler	32	7
	Lark sparrow	55	20
Shiny cowbird	Yellow-shouldered blackbird	38	18
	Rufous-collared sparrow	44	12
	Rufous-collared sparrow	30	19
Giant cowbird	Wagler's oropendola and yellow-rumped cacique (trees *with* wasps)	46	25
	Wagler's oropendola and yellow-rumped cacique (trees *without* wasps)	18	45

for the Kirtland's warbler and yellow-shouldered blackbird. There is little evidence that populations of other host species have been seriously affected. The reason for this apparent paradox is that a reduction in the number of young fledged by a host population does not necessarily result in a lower number of birds later on. If fewer young are produced, there may be lower mortality during periods of winter food shortages, so that similar numbers survive until the next breeding season.

The existence of brood parasitism raises a number of interesting questions. First, how did the habit get started? What were the conditions that favored the first depositions of eggs in the nests of other birds? Second, what made the habit spread? Third, what are the subsequent changes in the behavior of both the parasite and its host? Do hosts evolve resistance to parasites and do parasites evolve counter moves to those of their hosts? Fourth, how do young individuals of a parasitic species know to which species they belong? Fifth, do hosts and parasites communicate with one another?

How Did Brood Parasitism Get Started?

The origins of brood parasitism cannot be studied directly because nobody was around to witness the process, and it is unlikely that the evolution of a new parasite will occur rapidly enough for us to measure the alterations in its behavior. Nonetheless, there are clues that can be gleaned from the current behavior of brood parasites that may help us speculate about the most probable processes. Since brood parasitism has evolved a number of times among birds, the conditions favoring it are probably not highly restrictive.

It is not unusual for over half of the nests of small birds to be destroyed by predators before the young have fledged. Many of these losses occur during incubation; some occur while eggs are being laid. Since birds usually lay one egg each day and the process of forming an egg takes several days, a female whose nest is destroyed during laying will probably deposit somewhere one or two eggs that are already partly developed in her ovary before she reverts to the physiological and behavioral stages appropriate to initiation of a new nest. We do not know what happens to such eggs, but our knowledge of the ways in which behavior patterns are triggered suggests that the females probably seek out other nearby nests and lay their eggs there rather than simply dropping them on the ground. This is probably why close study of a species is likely to reveal sporadic cases of egg-laying in nests of other individuals, as we found among redwings.

In fact, since occasional laying of eggs in other nests probably occurs regularly in many species, we must ask why brood parasitism is not more common than it is. The relative rarity of the trait suggests either that it does not originate from occasional egg-laying by females whose nests have been destroyed, or that only rarely is the success of such dumped eggs high enough that females are favored to continue laying eggs in other nests to the exclusion of caring for their own offspring.

7.2 *A female giant cowbird inspects a nest of an oropendola to determine whether the time is appropriate to deposit an egg.*

7.3 Brown-headed cowbirds feeding near the head of a cow, the best place to capture insects disturbed by the movements of the large mammal.

What Conditions Helped Brood Parasitism Spread?

There are several ways in which individuals whose nests have been destroyed might do better by laying their eggs in the nests of other individuals rather than by dropping them on the ground. First, they would have greater reproductive success even if only a small fraction of the deposited eggs resulted in surviving offspring. Also, depositing eggs in the nests of other individuals costs relatively little and does not delay the time during which the individual can become ready to start its own replacement nest.

Even so, for brood parasitism to evolve fully, individuals that attempt to continue laying eggs in the nests of others must leave more surviving offspring than individuals that immediately initiate new nests of their own. Conditions for this to occur may be much rarer than those favoring occasional egg-dumping by individuals whose nests have been destroyed. We can infer what those conditions might be by observing the ecologies of current brood parasites.

All brood parasites except the black-headed duck are altricial, that is, they produce helpless young that must be fed for a considerable period of time before they are able to gather their own food. This suggests that parasitizing the food-gathering abilities of host parents is more important than parasitizing their egg-covering abilities. This could be especially advantageous to individuals of species in which food is hard for the adults to gather during the breeding season. Cowbirds, for example, are highly dependent upon disturbance by large mammals to create good foraging conditions (fig. 7.3). Today these conditions are found around domestic livestock and in agricultural areas. Cowbirds have recently expanded their range in the Sierra Nevada of California and now breed at much higher elevations than they formerly did. In these new localities the adults feed almost exclusively around horse corrals and daily fly long distances from those sites to the streamside areas where their hosts' nests are found. The general spread of the brown-headed cowbird in North America during the present century has been caused almost entirely by the creation of good foraging conditions for adults and not by changes in the availability of hosts. In areas into which cowbirds have recently spread, they use as hosts species that have been common there for as long as ornithological observations have been made, and, probably, much longer than that. Ancestral cowbirds may have had difficulty following wild grazing mammals during the breeding season while attending to fixed nests. They might have been forced to gather food for their young under conditions for which they were poorly adapted. This could have favored the use of the food-finding abilities of other species. Cowbirds, however, are the only brood parasites that forage around grazing mammals, and all other species that do so raise their own young. Unless we can uncover more definitive evidence, this speculation is not very convincing.

The evolution of brood parasitism results in the complete loss of nest-building, incubating, and feeding young in parasitic species. Whether the loss occurs as the result of a sudden genetic change that causes its bearers to cease all normal parental behavior and become immediate, full-fledged brood parasites or whether it evolves more gradually, beginning with occasional depositions of eggs in the nests of other individuals followed by a slow increase in that behavior at the expense of normal parental care, is not certain. There are reasons for favoring a gradual change over an abrupt one. An abrupt conversion would require simultaneous changes in a large number of physiological and behavioral patterns. The highly refined nest-searching behaviors of brood parasites are unlikely to appear suddenly. The success of eggs deposited by the first parasites was probably much lower than the success we can observe today, because most eggs would probably have been laid in nests where incubation was already underway. Such eggs are unlikely to hatch, and, even if they did hatch, the proportion that resulted in surviving adults would be low because late-hatched individuals are at a serious disadvantage.

Therefore, it is more likely that brood parasitism began with occasional deposition of eggs in the nests of other individuals by females who subsequently built their own nests and incubated a clutch of eggs. These females might, under some conditions, produce more surviving offspring than females who did not deposit any eggs in the nests of other birds.

Natural selection would favor deposition of more eggs in other nests at the expense of starting one's own nest if the average success of eggs deposited in other nests were higher. This success could be caused by lower predation rates on hosts nests, reduced competition with nest-mates, especially if hosts were smaller than the incipient parasite, and better foraging success of the hosts compared to the parasite. If such a process were to continue for a long enough time, a complete parasite that no longer attempted to build its own nest could evolve.

Brood parasitism might have evolved from a stage in which a species did not build its own nests but rather took over the nests of other species. This idea is appealing because some close relatives of brood parasites behave that way: bay-winged cowbirds and troupials, for example. It is tempting to see these species as representing stages—half-way houses as it were—on the road to full-scale parasitism. Nonetheless, taking over other nests does not obviously lead to the deposition of eggs in the nests of other individuals—the major change that must occur in the evolution of brood parasitism. Species that take over other birds' nests may be more likely to deposit eggs in the nests of other birds than are individuals of species that build their own nests, but there is no direct evidence of this. Taking over the nests of other species is favored when eggs and nestlings survive better in those nests than in nests built by the parents, or when the parents gain in saving time and energy by not having to build their own. This would be especially important when unusually sturdy nests are available. The bay-winged cowbird, for example, usually takes over large, domed nests of ovenbirds (fig. 7.4). These nests are safer than smaller, more open nests, and because of their sturdy construction large numbers of them persist for many years. These nests are generally very conspicuous, but it does not pay predators to visit all of them because many are unoccupied. Many small predators probably cannot break open those nests to take the eggs and young.

How Does the Behavior of Hosts and Parasites Evolve?

Because brood parasites usually reduce the nesting success of their hosts, the hosts should evolve behaviors that reduce the detrimental effects of the parasites. The best response would be to remove the eggs of the parasite and continue incubating one's own eggs. This may sound like a simple behavioral act, but it is really quite complicated. First, the host must be able to recognize its own eggs and to select the parasite egg for removal. Second, the host must be large enough to physically lift the egg from its nest. If the host is smaller than the parasite it may not be able to do so.

Since *we* can readily tells eggs of parasites from those of their hosts, the discrimination should be possible, but we should not assume that the task is as easy for birds as it is for us (fig. 7.5). Identifying parasite eggs may be difficult because animals seldom have the capacity to make discriminations that have not been important during their evolutionary history. Eggs do

7.4 Four bay-winged cowbirds inspect an old nest of the rufous cachalote, one of the large ovenbirds of South America.

not normally move around. In the absence of brood parasites or students of animal behavior, strange eggs do not appear in nests. Therefore, birds rarely have to distinguish their own eggs from those of other individuals. One's own eggs are those found in the nest where one laid them. A simple but striking example illustrates the point. Herring gulls lay speckled eggs and human observers can readily distinguish the eggs of nearby females. The gulls themselves, however, apparently cannot do so. If a gull's egg is placed on the rim of its nest, the adult will gently roll it into the nest and continue to incubate it. It does not matter if the egg is its own or one from another gull. However, if the egg is moved a bit farther from the nest, the gull will peck it open and eat it whether or not it is its own egg.

Although young gulls look very much alike to us, and we find them more difficult to tell apart than eggs, adult gulls readily distinguish their own young from the young of their neighbors. They feed and care for their own but attack and may eat the young of neighbors if they stray into their territory. Why are gulls so much better at telling chicks apart than eggs? Because chicks normally move around and it is important that adults feed their own young and not those of others. Nestlings of altricial birds do not move around, and adults are apparently unable to discriminate parasite nestlings and fledglings from their own (fig. 7.6). Rejection of young cowbirds by their hosts has not yet been recorded. As most potential hosts are probably unable to distinguish their own eggs from those of a parasite at the time parasitism first begins, this gives the parasite an initial advantage.

Still another problem for a host is that birds do not normally gain any advantage from tossing eggs out of their nests. Prior to the arrival of a parasite any bird that tossed eggs out of its nest would almost certainly be destroying its own eggs. To evolve the egg-tossing habit requires not only the ability to distinguish them but the evolution of very different responses to eggs from those of most birds. The genetic changes that produce such a marked departure from normal behavior may not evolve quickly.

Many potential hosts, however, do reject the eggs of parasites. This was not discovered until quite recently because it is very difficult to detect even with careful observations. If we regularly find parasite eggs in the nests of a host species, we can be certain that individuals of that species accept parasite eggs. But if we do not find parasite eggs, we cannot know if the parasites did not lay there in the first place or if the host removed them. It was not until Stephen Rothstein mass-produced artificial cowbird eggs made of plaster of Paris that the true answer was uncovered. Rothstein found that North American song birds could be divided into two groups—acceptors and rejectors. Most

7.5 Nest of a brown towhee in California with two towhee eggs and one cowbird egg (to the right).

individuals of rejector species, such as catbirds and Bullock's orioles, removed artificial cowbird eggs within a few days while most individuals of acceptor species never removed them. No species he tested fell in the middle.

Another response to the presence of a parasite egg is nest desertion or building a new floor to the nest that buries the parasite eggs together with those of the host. The host then lays a new clutch on top. This response is less effective than removal of the parasite egg because the host must abandon some of its own eggs and start over again. There is no guarantee that more parasite eggs will not be deposited in a relined nest or a new nest. Nonetheless, for some small North American song birds, this is the most common response to parasitism by cowbirds. These individuals may simply be too small to remove the cowbird eggs with their bills.

Rothstein argues that there are no intermediates between rejector and acceptor species because, once the appropriate behavior arises in some individuals of a species, the response spreads very rapidly through the population. The speed of the change makes it unlikely that we will find a species in the process of changing from its initial status as an acceptor to becoming a rejector. That rejection behavior should spread rapidly is obvious from the data in table 7.2. With such large differences in the success of parasitized and non-parasitized nests, rejectors have a great advantage over acceptors. Surprisingly, most blackbirds are acceptors, even though they are large enough to remove cowbird eggs. Three very different hypotheses have been

7.6 A red-eyed vireo feeds a fledgling brown-headed cowbird it
has raised, apparently oblivious to the fact that the cowbird
does not resemble a young vireo and is actually larger than the
adult vireo itself.

proposed to explain the persistence of acceptors.

One suggestion is that many species have only recently been parasitized by cowbirds. Given the close relationships between cowbirds and some of their hosts, this is unlikely to be the entire story. Another view, advanced by Rothstein, argues that rejection behavior, for the reasons discussed above, can evolve only with great difficulty. In many species, the right combinations of genetic changes may not have arisen even though they have been subjected to cowbird parasitism for a long time. As soon as the right changes occur, these species too will rapidly become rejectors.

A third view has been advanced by Amotz Zahavi, who notes that parasites may revisit host nests to assess the success of their offspring. Suppose a parasite, on finding that its eggs had been removed by the host, were to destroy all of the host's eggs. If parasites were efficient in relocating the nests in which they had already laid eggs, they could reduce the success of rejectors below the level of acceptors who do raise some young of their own.

Zahavi's hypothesis can be tested directly by the simple procedure of finding a large number of host nests in which parasite eggs have been laid and removing the parasite eggs from one group of nests and leaving them in another group. If the parasites are holding their hosts hostage, there should be a higher rate of destruction of nests in which the parasite eggs were removed than in nests where they were left. Preliminary tests of this type have recently been performed with red-winged and yellow-headed blackbirds, both acceptors, with negative results. But this does not reject the hypothesis for other host/parasite combinations.

Destruction of host nests is only one of the potential responses of parasites. If host individuals evolve the ability to discriminate between their own eggs and those of the parasite, parasite eggs that closely resemble those of their hosts are less likely to be removed than eggs that differ from those of their hosts. Because of this, egg mimicry has evolved in some parasites. Egg mimicry is best developed in some of the parasitic cuckoos of the Old World where individual females have eggs that closely mimic the eggs of a particular host species which they parasitize to the exclusion of all other species. In a given region there may be many "races" of cuckoos, each laying eggs of different appearances in the nests of a different host species.

Egg mimicry evolves slowly, and it is not yet well developed among the parasitic cowbirds, perhaps indicating that parasitism is more recent in this group than among cuckoos. Egg mimicry may be just beginning in the shiny cowbird. In October and November 1973, I studied marsh-nesting blackbirds near Pinamar, Argentina. Three species nested in my study area, two

of which were regularly parasitized by cowbirds. Of twenty nests I found of the yellow-winged blackbird, six contained one cowbird egg each. Of fifty-one nests I found of the brown-and-yellow marshbird, ten had one cowbird egg each and two had two each. Interestingly, all six of the eggs in yellow-wing nests were pure white and unspotted, while thirteen of the fourteen eggs I found in brown-and-yellow nests were spotted. Such a distribution is very unlikely to occur by chance, but could result if one or more females that preferred yellow-wing nests had laid all the white eggs. The white eggs are more similar to the eggs of the yellow-wing while the spotted cowbird eggs more closely resemble those of the brown-and-yellows. Unfortunately, my data do not distinguish between the possibility of different nest choices by female cowbirds laying different types of eggs and the hypothesis that the host species selectively removed eggs that differed more strikingly from their own.

A more difficult problem is to determine whether the parasites learn which hosts reject their eggs and if they then avoid laying in those nests. The use of artificial eggs does not help us with this problem. The only method is to find and monitor very closely many nests of a number of acceptor and rejector species. Some future ornithologist may test this interesting possibility.

Very little is known about how birds recognize foreign eggs in their nests. One possible mechanism is that they have an innate recognition of cowbird eggs but accept other eggs that are dissimilar to cowbird eggs. This is not the case with Bullock's orioles because they also reject eggs of loggerhead shrikes, house finches, and Brewer's blackbirds—none of which resemble cowbird eggs very closely. By putting cowbird eggs in oriole nests at different times in the building-laying cycle, Steve Rothstein found that Bullock's orioles learn their own egg type during a critical period that begins several days before the start of egg laying and ends the day the first egg is laid. Cowbird eggs placed in nests of orioles breeding for the first time are usually accepted while eggs placed in nests a bit later are rejected (fig. 7.7). Also, once an oriole has learned its own egg type it remembers it for the rest of its life. Cowbird eggs are rejected by orioles breeding for at least the second time no matter when they are placed there. As cowbirds do not normally lay eggs in a host nest until after the clutch has been started, this critical learning period normally serves orioles well.

How Do Young Parasites Know to Which Species They Belong?

A young cowbird, during its entire nestling life and for some time after it leaves its nest, finds itself in the company of individuals, both young and adult, of other

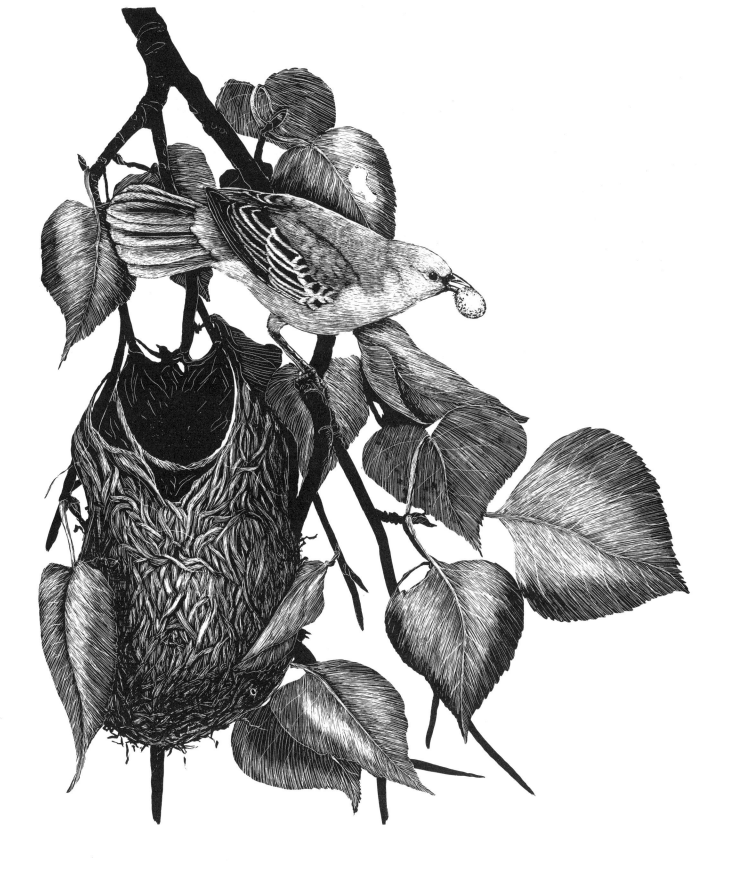

7.7 A female Bullock's oriole removes a cowbird egg from her
nest by piercing it with her bill.

*7.8 A male brown-headed cowbird performs a preening
invitation display in front of a female red-winged blackbird.*

species. Yet if it is to reproduce successfully it must
eventually reject the company of those individuals and
associate with cowbirds.But how does it know that it is
a cowbird or what cowbirds look like? Why doesn't it
develop an attachment to the species that raised it? Do
cowbirds have the avian equivalent of an identity crisis?

The question is interesting because we have known
for a long time that the young of some species of
precocial birds, such as ducks, geese, and chickens,
develop attachments to individuals of whatever species
they are associated with when they hatch. They may
even attach to humans and court them rather than
members of their own species when they become
sexually mature. Brood parasitism could not evolve if
those individuals successfully raised in the nests of
other species courted host individuals rather than
conspecifics. Therefore, from the beginning, species
recognition must not have been based on looking at
one's early associates. Almost nothing is known about
patterns of species recognition in altricial birds,
primarily because it is very difficult to determine how
they develop their sense of self-identity.

Species-identification mechanisms have been
investigated experimentally with captive brown-headed
cowbirds. Andrew King of Cornell University and
Meredith West of the University of North Carolina
discovered that a captive female cowbird, reared in the
laboratory in complete visual and vocal isolation from
other cowbirds, adopted a copulatory response posture
when they played the song of a male brown-headed
cowbird to her, but she did not respond in any way to
the song of a male red-winged blackbird. They then
took six female cowbirds, who were hatched by barn
swallows and fed by them until they were taken to the

laboratory, and hand-reared them as a group without
contact with adult cowbirds. Between the ages of thirty-
five and sixty days they were put into soundproof
chambers in pairs. Two were paired with other female
cowbirds, one with a female red-winged blackbird,and
one with a male cowbird. When the birds were nine
months old, recordings of male cowbirds, red-winged
blackbirds, eastern meadowlarks,and Baltimore orioles
were played to them. Those females that responded did
so more strongly to the songs of the male cowbird than
to the other songs. In the next phase of the test the
same songs were played to them, with the addition of
abnormal songs of a hand-reared male cowbird. During
that series, four of the females still showed a strong
preference to the abnormal songs of the hand-reared
male. The experiments showed that these cowbirds did
not require experience with adult cowbirds in order to
give species-specific responses.

Do Hosts and Parasites Communicate with One Another?

Many potential hosts recognize parasites near their
nests and attempt to drive them away. This is a form of
communication, and it may be effective in achieving its
ends. More subtle communication may also occur
between parasites and hosts.

In the laboratory, cowbirds approach individuals of
other species and give a preening invitation display in
which they step in front of another individual, turn
their head down, and fluff their feathers (fig. 7.8). They
are often preened by those individuals. The discoverers
of this behavior, Robert Selander and C. J. LaRue,
thought that cowbirds never preened one another or

individuals of other species. Therefore, Selander and LaRue named this display the "interspecific preening invitation display." They also postulated that the display functioned to reduce the aggression of potential hosts and enabled the cowbirds to gain access to their nests more easily than if they did not use such a display. A similar display has since been observed in four of the other five species of cowbirds.

Contrary to the original results of Selander and LaRue, Rothstein later found that cowbirds often displayed to and preened one another. Individuals most often display to members of the opposite sex. Rothstein also found that the frequency of displaying decreases with the length of time individuals have been together —which may account for the failure of Selander and LaRue to observe intraspecific displaying since their birds had been together in the same aviary for a long time. The preening invitation display may also have an important function in social interactions among cowbirds themselves.

The display is especially interesting because individuals of species that do not exhibit intraspecific preening respond by preening cowbirds. Cowbirds may use the display to test the willingness of individuals of the other species to accept their offspring, but until we know more about interactions between cowbirds and their hosts, we can only speculate.

Although ornithologists have learned a great deal about which species are parasitized by which parasites, about the evolution of egg mimicry, and about the effects of parasites on the reproductive success of host nests, there are many questions for which no answers exist. Why did brood parasitism evolve in the cowbirds and in no other group of New World song birds? Why is there but a single species of cowbird over most of North America? Why do some species of cowbirds parasitize more than two hundred hosts while others are confined to a single species or just a few species? Why do the cowbirds seem to prefer to parasitize other members of the blackbird family rather than more distantly related species? Clearly, much exciting research remains to be done on these fascinating birds.

8.1 A male melodious blackbird stands guard while his mate incubates her eggs in a nest slung near the trunk of a Caribbean pine.

8. The Roles of the Sexes

It is rare for males and females of a species to behave in the same way during the breeding season or engage in identical activities. This is as true for blackbirds as it is for people. Moreover, the factors that help and hinder individuals of the two sexes are usually not the same. Females sometimes benefit by changes in the social system that are not helpful to their mates and *vice versa*.

Conflict between the sexes has its roots in the fundamental difference between males and females. Ignoring for the moment the many complex and exciting differences between the sexes, there is one key difference from which all of these other variations have sprung. By definition, a female is an individual that produces sex cells (eggs) endowed with a significant energy reserve while a male is an individual that produces sex cells (sperm) with very little energy reserve, typically just enough to allow them to swim to the eggs. Whenever biologists find a difference in the size and energy reserves of the sex cells, they speak of maleness and femaleness. When they find different types of sex cells that are of approximately the same size, they refer to mating types, but not males and females.

This simple difference has profound evolutionary implications. Because of these differences the number of eggs that can be produced by a female is normally limited by the amount of energy she can accumulate. The production of sperm, by contrast, is cheap; large numbers can be produced at low cost. The result is that in most species the reproductive success of a female is little influenced by the number of males with whom she mates. Rather, it depends on the quality of the genes the male contributes and his role in providing parental care. A male, in contrast, can typically increase his reproductive success by mating with more than one female, whether or not he contributes parental care to the offspring resulting from most of those matings. Females, because of their large reproductive energy commitments, are a limited resource for which males compete.

Since males usually compete for females and a single male may inseminate many females, reproductive success is usually more variable among males than it is among females. Some males may achieve no fertilizations while others fertilize the eggs of many females. There is never that much variability in the reproductive success of females. The Guinness Book of Records shows that Movlay Ismail, an Emperor of Morocco, sired 888 children while the record human female

produced 69 offspring in 27 pregnancies! Many male contemporaries of Movlay Ismail sired no offspring at all. Most females of the day delivered children.

Nonetheless, whatever the intrasexual variability in reproductive success, half of the genes in each generation come from males and half from females, no matter what the mating pattern. If there are approximately equal numbers of males and females, the average *per capita* genetic contribution of each sex is about the same. That is, the average male contributes as much as the average female to the genes of the next generation even if one male inseminates all of the females in the population. Mating patterns are never that extreme among blackbirds, but a male redwing with a good territory may attract over a dozen females. For every such male, many others go without mates.

These basic genetic facts go a long way toward explaining why individuals of opposite sexes usually behave differently during the breeding season, but the importance of these facts has only recently been appreciated by evolutionary biologists. The frills of sexual differences had diverted attention away from the essence of sexual differences, which is that more energy is devoted to eggs than to sperm.

Although there are differences in the energy contents of eggs and sperm, the sex cells themselves constitute only a small fraction of the total energy expended by birds in reproduction. Individuals may defend territories, court mates, drive away predators, and care for eggs and young. In combination these activities dwarf the energy expended in making eggs and sperm. Care and tending of eggs and young reduce the ability of the parents to produce offspring in the future. The longer current offspring are cared for the greater the time before the parents can produce another set, the more weight the adults are likely to lose, and the more at risk to their predators they will be. Many breeding-season activities, such as territorial defense, are not directed specifically toward particular current offspring, and they do not necessarily affect opportunities for future reproduction.

A male usually gains genetically by devoting time to attracting and copulating with additional females, while a female benefits if a male devotes all of his attention to herself and her offspring. One partner may gain by deserting its mate and investing in another set of offspring with another individual. Who deserts first depends in part on who has the better opportunities elsewhere. Among birds this is often, but not always, the male.

Other than the acts of producing and fertilizing eggs, done by females and males respectively, there are no tasks during the breeding season that cannot be performed with equal ease and effectiveness by both male and female birds. This is in marked contrast to mammals where only females possess mammary glands for the nourishment of the young and males are, accordingly, excluded from an important part of breeding responsibilities. The marked differences in sexual roles that occur in birds are caused by factors other than the abilities of individuals of one of the sexes to perform those acts. Rather they are due to evolutionary changes favored because individuals of one sex increased their genetic contributions to future generations by dropping or accentuating particular types of behavior. One of the most interesting areas of biological research is finding out why different patterns of behavior have been successful in different species.

The Costs and Benefits of Performing Different Activities

In the world of behavior, as in the world of economics, one does not get something for nothing. Free lunches are universally a scarce commodity. All components of reproductive effort have a time, energy, or survival price-tag on them. Other activities that might be engaged in must be forgone. As with our own lives, the difficult problems faced by blackbirds involve choices among beneficial activities, not all of which can be performed in the limited time available. What an individual does also depends on what its mate does, and *vice versa*. Important decisions must often be made on the basis of very inadequate information. How good will the territory be in a month's time? How reliable will a potential partner be in caring for offspring? Information is valuable but often difficult to obtain. Social organization is not an arena for individuals who like things nice and tidy and want a secure vision of the future.

The key social decisions made by a bird during the breeding season are its choice of breeding site, its choice of mate, the investments it will make in the pair bond, the care it will deliver its offspring, and when it will desert the young, its partner, or both. Although choice of partner is made early, it is more convenient to treat that decision last because we need to know what the individual and its mate do before we can say anything useful about how partners ought to be picked.

In an earlier chapter we noted the striking rarity of nest construction by male blackbirds. Here we only need to integrate that information with other breeding season activities.

Incubation

In most families and orders of birds, both sexes share more or less evenly in the task of incubation. Sometimes one sex incubates primarily at night while the other does so during the day. In other cases, the members of the pair alternate on the eggs at more frequent intervals. In some groups incubation is performed entirely by individuals of one sex. The hummingbirds are striking examples—incubation is performed strictly by the female in every species thus far studied.

Despite their variability in almost every other feature of their social organization, blackbirds are apparently absolutely constant with respect to this trait. Only females incubate. Even among the tropical orioles where males and females are equally bright, and both sing, males have never been reported to incubate. Admittedly, information is sparse for many species, but, from what we know already, incubation by males must be a rarity in the entire family. This is in contrast to the situation in closely related families of birds, such as finches, where males of many species incubate regularly.

Even though male icterids do not incubate, they are still attentive during this phase of the breeding cycle. Males of many species perch near the nests and warn females of possible danger. This is the case with monogamous species such as the melodious blackbird, common grackle, and Brewer's blackbird, and with polygynous species such as redwings, yellowheads, oropendolas, caciques, and great-tailed grackles (fig. 8.1). Several factors may make a guarding role a preferable activity for males of polygynous species. First, a male can simultaneously guard one female *and* attract additional females. Second, in these species males are usually much larger than females and, hence, are better adapted for defending the nest. Finally, if males are brighter than females, they may be more likely to reveal the location of the nest by incubating instead of sitting at a distance and giving information to the female. Nonetheless, among monogamous species the sexes are often nearly alike in size and plumage, making individuals of both sexes about equally adapted to perform all of those activities. Yet the striking sexual difference persists.

It is easy to study nest defense behavior by placing live or mounted predators near active nests and observing the responses, but these experiments cannot tell us about the effectiveness of the behavior because the predator is not free either to kill or to leave. Despite these limitations, such experiments tell us a great deal about the vigor with which individuals of both sexes attack and how their behavior varies with the stage of the breeding cycle. We would expect, for example, birds to defend newly constructed nests less than nests with eggs, and nests with young still more vigorously, because the value of the nest increases with the age of its contents. Young nearly ready to fledge have already survived many days of possible mortality, both as eggs and as younger nestlings. Their probability of surviving to adulthood is much higher than that of an egg or younger nestling. The loss of older offspring is more serious than the loss of younger ones.

Males of polygynous species have less at stake with the success of any single nest than does the female owner of that nest. In a monogamous species, however, the loss is about equal for both male and female if the nest is destroyed. Therefore males and females should attack predators with equal vigor in monogamous species but females should attack more vigorously in polygynous ones.

These predictions were tested with blackbirds at the Potholes by Eleonora D'Arms, who used live black-billed magpies and goshawks as predators. These two birds differ in the type of threat they pose to blackbirds. Magpies attack only eggs and nestlings while goshawks mostly attack adults and fledglings. Predators were tethered near nests for fifteen minute periods, and the number of vocalizations, flights toward the predator, and actual physical attacks were recorded. Predators near finished, but still empty, nests were largely ignored, but once eggs were laid mobbing behavior became prominent. Females increased their mobbing effort once the eggs hatched, but overall effort of the males did not change at hatching. Females with nestlings mobbed more vigorously than females with eggs, flying at and striking the magpie much more often (fig. 8.2). (These attacks, though often vigorous, seldom deter magpies from eating eggs and nestlings.)

In these kinds of tests there are large individual differences. Some males attack predators very vigorously while others merely sit at a distance and call. A few birds actually disappear temporarily from their territories during a test. This variability is not simply an artifact of the test situation because it can be seen in nontest situations as well. I can well remember certain male redwings whose territories I always entered with great caution. They invariably attacked me from behind, hitting me on the head hard enough to cause discomfort. At other territories males did nothing more than call from a distance. Sievert Rohwer and his co-workers have found that there is a strong correlation between the vigor with which male redwings attack predators and the strength of their attacks against other male redwings. Males possess something like a general aggressive tendency that affects their interactions with both competitors and predators.

While males are guarding nests, they may also continue to attract additional females. The frequency with which male redwings perform displays that serve to attract females does not diminish when the male acquires his first mate, or even subsequent mates. By perching, singing, and patrolling the territory he can both attract more females and look out for predators.

Females may increase their success by repelling additional females while they are building their nests and incubating, but it is much more difficult for them to do so if they perform all of the incubation. An incubating female must get off her nest to chase intruders, thereby leaving her eggs to cool. A persistent intruder can easily force an incubating female to give up her attempts to defend her territory because letting another female settle is not as serious as losing the eggs or young she already has.

Prior to incubation females may devote a great deal of effort to defense of territories. In the Potholes, the first female redwings to settle often defend all of their mates' territories (fig. 8.3). These are typically old, experienced females, and many of them have bred in the same territory in previous years. Their defense may prevent additional females from settling until they actually begin to incubate. Later-arriving females do not engage in much of this type of behavior. Since there is considerable starvation among early nestlings but not among later ones, there is stronger reason to prevent settling of more females earlier in the season than later. As the season progresses, predation rates increase. In Washington State we have found a weak negative correlation between the number of nests in a territory and the probability that a predator takes a nest. Additional females do not pose a threat to already established females later in the season as they do earlier. The females are making rather subtle adjustments in their behavior according to the situation and are not engaging in fixed behavior patterns regardless of the circumstances.

Guarding the Female to Avoid Being Cuckolded

There is an old Chinese proverb that states, "It is a wise man who knows his own son." The deep wisdom of that statement derives from the fact that among animals with internal fertilization, such as birds and mammals, a male can never be certain that he is the father of his presumed children. By a variety of devices he can increase the probability that he is, but certainty can never be achieved. Females, on the other hand, face no such uncertainties.

The most effective way to avoid being cuckolded, and practiced by males of many species of birds, especially monogamous ones, is to follow their mates around very closely during those periods when they are copulating (fig. 8.4). In most birds this period begins about mid-

8.2 *A female red-winged blackbird attacks a magpie at her nest.*

way during the construction of the nest and continues through egg laying, a relatively brief period of about six to eight days. In some species a female copulates only once and fertilizes all her eggs from that single batch of sperm, but among song birds the norm is for a female to copulate several times each day.

In Argentina I often watched a male brown-and-yellow marshbird accompanying his mate as she gathered nesting material. The pair flew back to the nest area together and the male perched nearby while the female incorporated the material into the nest. Similar behavior characterizes common grackles and Brewer's blackbirds at the same stage in the breeding cycle. Once the nest is finished, an egg is deposited once each day, usually early in the morning, and females spend most of their time foraging. It is much easier for the males to follow them at this time.

Males of some species take advantage of the nest-guarding phase to attempt to inseminate females elsewhere. Although it is not easy to follow male blackbirds during that period, one way to infer their activity is to monitor the amount of time they spend away from the vicinity of the nest while the female is incubating. By watching the nest we can determine how much time it takes the female to feed herself because most of the time she is off the nest, particularly in cool weather, is devoted to feeding. If we further assume that a male can find the food he needs to maintain himself in slightly more time than the female needs, we can estimate the amount of time he is away but not feeding. This is time that could be devoted to courting and other similar activities. Male yellow-winged blackbirds in Argentina usually sit near their nests and sing occasionally while the females incubate, but they are away longer than they need to be to find food. Occasionally they can be seen courting other females at that time.

Feeding Nestlings

The young of most small- to medium-sized song birds grow to nearly adult size within two or three weeks of hatching. During part of this time they eat more than their own weight in food each day. During the middle and latter parts of the nestling period, the adults of a pair must provide each day several times the amount of food that would be needed for simple self-maintenance. If both adults feed the young, more food can be brought each day, allowing either faster growth of the young or a larger number of young to be raised. There are substantial losses if both parents do not feed, and any parent that fails to bring food should be engaged in some very profitable activity that more than compensates for the loss.

What are these other activities? One obvious candi-

8.3 A female red-winged blackbird uses a song spread to defend her territory within the larger space held by a male.

date is providing protection against predators. Among larger birds that nest in conspicuous places, nestlings below a certain size cannot be left unattended without exposing them to substantial risks. Unattended eggs and young of colonial sea birds are quickly eaten by gulls that constantly patrol through the colonies. Typically, one adult is in attendance and the other foraging most of the time. In none of these species do individuals of one sex fail to feed the young. They simply alternate their efforts.

Another possibility is that an individual could spend time attempting to attract additional mates or to seek extra copulations. Opportunities to do so are usually much better for males than they are for females. A female would have to find another male willing to incubate a new clutch of eggs and her first mate must be willing to continue caring for the first clutch. Otherwise she would reduce her reproductive output by abandoning her first brood. Males, however, can benefit from inseminating additional females even if

they do not help care for the young produced. Also, males of most temperate-zone birds defend territories, thereby preventing many other males from breeding. When this happens, and the sex ratio is approximately equal, there will be extra females to be attracted to the territories. A male may do better by spending his time courting and advertising for more females than by feeding his existing young.

Given these considerations we expect both males and females to feed the young in monogamous species and for both of them to invest about the same amount of effort in that task. Males of polygynous species have more to gain from alternative expenditures of time, so many of them should not feed young. Among polygynous species the amount of feeding of young should vary according to the chances that the male can attract more females if he continues to advertise and according to the value that his feeding will have for the survival of the young.

Males feed young in all of the monogamous species of blackbird that have been studied, but this is the case in only 40 percent of the polygynous species (table 8.1). If no male oropendolas feed young, as seems likely, the figure would be about 30 percent. Among the polygynous species there are some interesting patterns that tell us a great deal about the factors influencing male parental behavior. In many parts of eastern North America, where harems are relatively small, and the breeding season is relatively short, many male

redwings feed nestlings on their territories. In a well-studied population near Bloomington, Indiana, about half the males feed nestlings in at least one of the nests on their territory. In contrast, such behavior is very rare in the state of Washington, where only about 5 percent of males feed nestlings in most years.

Many male redwings in Washington do feed fledglings later in the season, however. At that time, opportunities to attract additional females have become relatively poor and males apparently gain more by feeding already produced young than by advertising for more mates. Owing to the short nesting season in Indiana, males reach the point where feeding young is a better option sooner in the season, many of them getting there when their young are still in the nest. This idea is supported by the behavior of male redwings in the Potholes in 1981. As a result of the eruption of Mount St. Helens on May 19, 1980, which covered our study area with a blanket of ash one inch thick, very few young redwings were produced in the Potholes that year. In 1981 fewer females than usual arrived late in the breeding season and many male redwings fed nestlings!

Most male yellow-headed blackbirds feed nestlings, concentrating their efforts at the first nest to hatch on their territory. Secondary and tertiary females can expect help from the male only if something happens to the first nests. Male yellowheads are also flexible in their behavior. David Gori has found that if a male

Table 8.1. **The Role of Males of Polygynous Icterids in Feeding Young (by number of species)**

Group	Males Feed Young	Males Do Not Feed Young
Oropendolas		4 (probably all 10)
Caciques		3
Yellow-headed blackbird	1	
Redwings and allies (Agelaius)	2	
Meadowlarks (weakly polygynous)	2	
Grackles		2
Bobolink	1	
Totals	6	9 (15)

yellowhead continues to attract more females he may not feed any of his nestlings. The larger his harem, the less likely a male is to to take time away from courting to feed young.

If males are responding to immediate conditions, we should be able to manipulate these conditions and change the behavior of the birds. We did this at the Potholes by manipulating clutch sizes. The normal clutch size of yellowheads is four eggs. With the full complement of young, some individuals grow more slowly than their nest mates and the last hatched young often starve. A male can make a substantial contribution to the survival of his offspring by feeding them when there are four nestlings. But if there are only two young in a nest, the value of his feeding is not as great because the female can provide for them by herself. In that case a male should switch his provisioning to his second nest if it has a full complement of young. We reduced clutches to two eggs in primary nests on twelve yellowhead territories. All these males switched and fed at their second nest. When we attempted to shift them to the third nest by reducing clutches in both the first and second nests, they stayed with their first nest. This is not always the case, however, because David Gori has been able to divert males to even later nests by the same technique. His experiments were run in a different year with males having larger harems than those used in the earlier studies. We do not yet know the full flexibility that male yellowheads can show in their parental behavior, but it is evident that males are able to monitor local conditions and change their behavior in ways that increase the number of young they produce during a breeding season.

When Should Birds Desert Their Young?

All young birds are eventually deserted by their parents because parental feeding stops at a time when additional food is probably still of some value. If the breeding season is long enough to permit another brood to be reared, parents desert their young sooner than they would if no new nest could be started that season. Often one sex deserts before the other. If males do not incubate eggs or feed their young, they have largely deserted the offspring as soon as the eggs are fertilized, but they may continue some investment by guarding.

Among monogamous birds male and female usually desert the young at about the same time. If another nest can be attempted, the female may desert first to build a new nest and accumulate energy for the next clutch of eggs, leaving the young to the care of the male. The same thing happens among polygynous species. Female redwings sometimes stop feeding fledglings and start a new nest while the male continues to bring food (fig. 8.5). Both sexes gain because they are able to produce a new clutch sooner than would otherwise be the case. Only if the female leaves her mate to start her new nest on a different territory is the male a true loser. Among redwings this usually happens only when the first nest is lost to predators and there are no young left in the care of the male.

How Females Choose Their Mates

Having considered the kinds of reproductive investments that males and females make during the

8.5 *A male redwing feeds a fledgling from his first brood while his mate builds the nest she will use for her second clutch of the season.*

breeding season, we are now in a position to return to their choice of spouse. The criteria used in making that decision depend on the role the spouse will play in the breeding effort. If the male provides no territory with nest sites or food resources, and he does not feed the young, female choice is based entirely on the genes the male will provide. If the male provides a territory within which food for the female and her offspring will be gathered, the quality of his real estate is a major factor in the choice of the female.

In monogamous species, individuals of both sexes have about the same stakes in choice of a partner. Both the male and the female are making a commitment for a full breeding cycle and both are closing other options for that period. Both male and female suffer equally if

they make a poor choice. Among polygynous species the male has less at stake in any one choice of partner than does the female. If he can attract several females, he should be less concerned about the quality of each one. If his first female greatly restricts access by other females, the male's concern with his first choice should go up, but it is unlikely to reach the same level as that of the female.

Certain conditions are necessary for a trait to be used in mate selection. First, there must be variability among individuals in that trait. Second, that variability must be perceivable by the choosing individual. Third, the variability must actually influence reproductive success. As an example, male redwings are highly variable in their aggression toward predators. Should females use that behavior in choosing a mate? Evidently not, for several reasons. First, it may be difficult to determine how a male actually responds to predators because none may come around during the courtship period. It may not pay a female to delay her choice to allow the appropriate situation to arise. Besides, Bill Searcy could find no correlation between his measures of aggressiveness of males toward predators and other male redwings and the percentage of nests on their territories that were actually taken by predators. Our third criterion was not met—there were no measurable effects of that behavior on reproductive success. Predation rates depend strongly upon the position of the territory and the quality of its nest sites but not on the behavior of the males.

These results raise another puzzling problem. If anti-predator behavior really has no influence on the activities of predators, why do redwings spend so much time engaging in it? Are they wasting their time? Since that seems unlikely, particularly since they expose themselves to danger by actually attacking dangerous predators, there must be other explanations. One possibility is that some males are more aggressive toward predators *because* their territories have poorer nest sites than those on territories of less aggressive males. If so, predation rates might have been still higher on those territories if their owners had not been vigorous in their attacks on predators. Similarly, males with good-quality territories might be less aggressive because they do not have to be. Such site-related variability would obscure relationships between aggression and predation rates. It is very difficult to test this idea. The best way is to observe males over the course of several years during which time the quality of the nest sites on their territories change. If the males change their behavior accordingly, this would be strong evidence of the hypothesis. Unfortunately, males usually return to the same territory year after year and, unless there are changes in the same territory between seasons, an opportunity to make the test will not arise. This quality of information is not available for any species of blackbird, or for any other species of bird for that matter.

Although the evidence suggests that female redwings do not consider male aggression as one of their criteria, they are not totally insensitive to traits of the male. In a study of redwings breeding in the Fraser River Delta of southwestern British Columbia, Jaroslav Picman found that the size of a male's harem increased with the number of years he had held his territory. When an old male disappeared and a new male took over the territory, the number of females settling there dropped from what it had been the previous year. As a new male holds a territory over the years, his harem size increases. Bill Searcy found the same thing for

redwings at the Turnbull Refuge. That the male occupying the territory does make a difference in female choices is also suggested by the fact that males devote all of their time to advertising and do not feed their young while new females are still arriving in good numbers. If the male on the territory were irrelevant, females should settle whether he courted vigorously or not, and he would be better off feeding his young.

Some characteristics of the male must be important in redwings even though it is difficult to determine what they are. Bill Searcy examined many characteristics of males, including their sizes, size of their red epaulets, width of the yellow band below the red, and song rate, but of these variables only song rate correlated at all with harem sizes—and that only weakly. The criteria for mate selection by females remain very elusive.

Real estate is also very important to female redwings. Arriving females immediately drop into the vegetation and explore potential territories at the water surface, looking for insects and nest sites. They move rapidly among territories and may visit a large number of them in the process of deciding. To a human observer they seem to be totally oblivious of the males and their striking behavioral antics, but they probably notice more than we suspect.

When Males Are Choosy about Females

Males of monogamous species should be as particular in their choice of mates as are females, but we know very little about how individuals of those species select mates. Even male redwings are not receptive to all females. Males are sometimes aggressive toward the first females to arrive. These females commonly defend the entire territory from other females until they have built their nests and laid their eggs. Therefore, the male has a substantial interest in the quality of the first female. At this time during the breeding season male redwings chase females very vigorously (fig. 8.6). This behavior has been puzzling to ornithologists for a long time, but it may be a way of testing female quality. If she can fly very rapidly and has good maneuverability, she may be a better forager who will raise better-fed young.

Support for this idea comes from the observation that males chase later-arriving females much less. In fact, once the first female has laid her eggs, the male is more likely to intervene on behalf of a newcomer, attacking his first mate when she tries to prevent the new female from settling. Since these later females are never territorial and since nesting success remains high as harem sizes increase, a male does not benefit from being selective at that time, whereas he does benefit by being selective earlier in the season. The criteria by which he chooses are still unknown.

While there is much to be learned about sex roles among birds, it is already clear that birds are much more subtle and flexible in their responses than we had expected. The earlier literature often referred to a species as being territorial, or stated that males of a particular species always or never did certain things. We now realize that most of these generalizations are too simple and that individuals change their behavior in accordance with local circumstances. This opens up very exciting possibilities for research because if behavior is flexible, then it can be studied by manipulations in the field. The few field manipulations of sex roles that have been performed in recent years are certain to be followed by a veritable flood of studies designed to determine how flexible individuals are and to which factors they are most responsive.

Size and Color Differences between the Sexes

If males and females use different criteria in choosing mates, and if individuals of the two sexes are engaged in different activities during the breeding season, it is not surprising that they have evolved different sizes and colorations too.

If individuals of both sexes perform all activities (other than egg and sperm production) during the breeding season, the same size should be best for both of them. Among blackbirds, males apparently never incubate, but incubation is not an activity that should exert a strong influence on the sizes of birds. In monogamous species with nearly equal roles, there is little size dimorphism, usually less than 10 percent difference between the sexes.

Among migratory monogamous species there is an additional difference in male and female roles. The males typically come to the breeding grounds first and establish and defend territories. Larger size contributes to the defense of a good territory. In species where males, primarily or exclusively, defend the territories, they should be somewhat larger than the females.

How much larger depends on other consequences of getting bigger. A larger bird requires more food each day and has to spend more time getting it. Also, a larger bird may be less mobile and a more conspicuous and tempting target for predators. We have already noted that male great-tailed grackles die at about twice the rate of females during the winter, and that male cowbirds are more likely to die during cold weather than females.

It becomes apparent that the extent of size difference should depend on how both the costs and benefits increase with size. As long as benefits increase faster than costs, larger size is favored, but when costs begin to increase faster than benefits, selection for larger size

stops. Among monogamous species the main advantage of getting a better territory is that, on the average, more young can be reared during a breeding season. But a monogamous male attracts only one female and most males with poorer-quality territories also get females.

A male of a polygynous species can attract a number of females, each of which may be about as successful in rearing young as any other. The expected payoff of holding a good territory increases with harem sizes. The larger the average harem, the greater should be the size differences between males and females.

These predictions are easily tested because information on size is available for all species of blackbirds (table 8.2). Monogamous species in which both sexes defend the territory show very little sexual size dimorphism (fig. 8.7). Males and females are within 5 percent of one another in size. Among monogamous species, in which males are the primary defenders of the territories, males average about 10 percent larger than the females, while among polygynous species males average about 20 percent larger than females. The largest size dimorphisms in the family occur among the densely colonial oropendolas and tropical grackles, where males may be 35 percent larger than females and weigh twice as much (fig. 8.8).

Role differences during the breeding season also favor different color patterns. The more similar the roles of the two sexes the more similar their plumages should be. If males play a greater role in defense of territories, they should be correspondingly brighter than their spouses. Defenders of harems should be brightest of all.

Tropical icterids, in which both sexes defend the territory, are monomorphic in color. In some species males can be distinguished from females only by their behavior or examination of their internal organs. In other species, the sexes can be distinguished by their appearances, but the differences are very slight. In a few temperate species, such as the scarlet-headed blackbird, pairs are formed before territories are established and the pair arrives together on the breeding area and establishes and defends its territory together. In these species, too, the sexes are very similar.

There is not, however, a strong correlation between the extent of polygyny and the amount of sexual differences in color. Some of the most strongly polygynous species, such as great-tailed and boat-tailed grackles, oropendolas, and caciques, though being quite different in size, are not markedly different in plumage. Males are readily distinguished from females in all these species, but they are no more different from females than males of most monogamous species. In fact, some of the most striking sexual dimorphisms among the blackbirds occur in the migratory,

8.6 *A sexual chase in which several male redwings temporarily violate territorial boundaries.*

Table 8.2. **Mating Systems and Size Dimorphism Among Icterids**

Group	Mating Pattern	Spacing Pattern	Average Difference in Size between Sexes (in percent)	Range of Size Differences (in percent)
Oropendolas	Polygynous	Colonial	25	15-35
Caciques	Polygynous	Colonial	22	21-23
Caciques	Monogamous	Territorial	12	10-15
Orioles	Monogamous	Territorial	6	0-14
Oriole blackbird	Monogamous	?	5	
Yellow-headed blackbird	Polygynous	Territorial	20	
Redwings and allies	Polygynous	Territorial	15	12-18
Other *Agelaius*	Monogamous	Varied	7	6-14
Meadowlarks	Monogamous	Territorial	8	5-11
Brown-and-yellow marshbird	Monogamous	Colonial	5	
Scarlet-headed blackbird	Monogamous	Territorial	6	
Austral blackbird	Monogamous	Territorial	6	
Melodious blackbird	Monogamous	Territorial	10	
Grackles	Monogamous	Territorial	13	11-14
Grackles	Polygynous	Territorial	21	20-22
Euphagus	Monogamous	Colonial	7	5-8
Small cowbirds	Monogamous	Varied	9	5-12
Giant cowbird	Polygynous	Colonial	24	
Bobolink	Polygynous	Territorial	11	

8.7 A pair of yellow-backed orioles defends its territory against an intruding pair. Although the sexes are almost identical in plumage, males attack primarily males, while females attack primarily females.

monogamous orioles of North America (fig. 8.9).

This pattern among blackbirds is unexpected because in many other groups of birds, such as pheasants, grouse, birds of paradise, cotingas, and manakins, there is a strong association between polygyny and sexual differences in coloration. Males of these birds all gather together on communal display grounds, or leks. Females come to the display grounds to choose their mates, and their selections (although we do not know how they are made in any of these species) must be based entirely on the characteristics of the males because none of them defend any resources. Many polygynous male icterids also defend no resources, but holding or not holding resources cannot be the main determinant of plumage dimorphism, because sexual differences in plumage are most marked in those species, such as redwings and yellowheads, in which the males *do* defend resources, and less marked in

species, such as oropendolas and colonial grackles, where they do not.

Many aspects of the sex roles of blackbirds remain puzzling. On the one hand there is evidence that individuals can adjust their behavior to immediate circumstances, shifting their efforts from nest to nest and adjusting the total amount of parental care they give. On the other hand, there is evidence of great inflexibility. Why do male icterids never incubate? It is difficult to believe that incubation by males is never advantageous. The failure of males of most blackbirds to bring food to incubating females is also puzzling. The problems may lie with our imaginations rather than with the adaptability of the blackbirds themselves. Alternatively, the patterns may indicate constraints on the pathways by which these birds adapt to their environments.

8.8 As is typical of polygynously breeding species, the male great-tailed grackle is much larger than the female.

8.9 Tropical resident orioles are all monomorphic in plumage, but among all species that migrate to the north temperate zone to breed, males are bright and females are dull. Shown here are the migratory hooded oriole that breeds in the southwestern United States and adjacent Mexico, and (facing page, top) the Atamira oriole of Mexico and (bottom) the epaulet oriole of South America, both of which are residents in their respective ranges.

9.1 *A male western meadowlark advertises his territory with a*
melodious song (sonagram of his song shown at right).

9. On Communication

Field guides often have little lines pointing to the plumage characters most useful in identifying the birds illustrated. One can make much better use of brief glimpses of birds in bushes if one knows just where to look to find the key mark that will provide unambiguous identifications. More recently, sonagrams of the songs of the birds have been included next to the illustrations, providing, for those who know how to read them, yet another set of key characters (fig. 9.1). Since *we* use these traits to tell one species from another, it was natural to assume that the *birds* used them for the same reason. This notion was compelling for another reason. For several decades biologists thought that the major challenge facing individuals of most species was to avoid mating with individuals of other species because such matings would either be infertile or would produce poorly adapted offspring. Biologists spent a great deal of time showing that each species had distinct communication signals by means of which every individual could be identified to its species.

More recently, however, it has become apparent that birds use communication signals and visual patterns to make much more subtle discriminations. A yellow-headed blackbird uses plumage patterns primarily to measure small but important differences among other yellowheads with whom it may forage, compete, fight, or mate, rather than to distinguish its conspecifics from individuals of other species. Differences in song types are important in mate selection and territorial behavior. Birds adjust their songs to those of their neighbors. Slight differences in color patterns signal dominance status. Birds do recognize the species to which an associate belongs by voice and plumage; but, to understand their signals, we must recognize that the messages they send and how these are interpreted are more complicated than signaling species identity.

Individuals of all social animals regularly communicate with one another. While on their territories, male redwings make some sound about six times every minute, about one-third of which are songs and two-thirds shorter sounds such as alarm, flight, and courting calls. About half of those vocalizations relate to defense of the territory. Redwings coming into their winter roosts also constantly sing and call even though we do not understand very much about what they are saying and why. Bird song has long been a source of inspiration for people. The haunting, repeated call of the European cuckoo and the nocturnal melodies of the nightingale have played a prominent role in European literature for centuries. We can gain a different but equally important inspiration from understanding where the signals came from, why they evolved, and what they communicate.

The Origins of Communication Signals

The first person to study communication signals systematically was none other than Charles Darwin himself. In 1873, he published a book entitled *The Expression of the Emotions in Man and Animals;* unlike his earlier book, this one attracted almost no attention. Most of Darwin's ideas were rediscovered nearly a hundred years later. One of his important observations was that animals, by their actions, often give clues about their future behavior. A bird about to take off may, by its preparatory movements, reveal its intentions to others around it. The physiological changes undergone by an individual preparing to fight may tell its associates what it is about to do. Darwin recognized that these involuntary behaviors, which provided information to associates, were the events from which many communication signals arose.

How Communication Signals Evolve

It is a long way from a flight intention movement in which a bird crouches, sleeks its feathers, and slightly raises its wings at the shoulders to the elaborate song spread of a male tricolored blackbird in which the wings are fully spread, the red feathers on the shoulders are erected, many of the body feathers are ruffled, and the bird utters a loud vocalization (fig. 9.2). Yet the latter probably evolved from the former.

To understand how and why we must remember that social systems are composed of individuals whose evolutionary interests overlap but do not coincide. Any member of a social group potentially can benefit by causing a change in the behavior of its associates. The way such changes are brought about is through signaling. The receiver of a signal, however, has its own interests, which may be different from those of the sender. It should respond to the signal in ways that improve its own fitness.

Like all other activities, signaling has a price tag. It costs something to have the ability to perform a signal,

9.2 *When giving a high intensity song spread, a male tricolored blackbird lowers and spreads his tail and wings, erects his epaulets, and raises nearly all of the feathers of his head and body.*

and it costs something to actually perform it. Giving a signal is likely to make the individual more conspicuous to its predators than if it were to remain quiet. Also, other individuals can respond to the signal by changing their own behavior in ways that might be detrimental to the signaler. What you say may actually hurt you!

Two important principles guide the evolution of communication signals. First, no signal can evolve if its receivers are able to respond in ways that, on average, hurt the signaler, unless receivers and signalers are close relatives. Second, no signal can evolve if there is nothing the receiver can do to improve its life when it receives the message. Otherwise the receiver is better off ignoring the signal, which can then never evolve into a true message. Singing is worthwhile only if other individuals listen.

The begging of nestling birds illustrates the principles. A young bird signals its hunger by uttering begging notes. These are perceived by its parents, who respond by bringing food. The nestling gains by being fed and the parents gain by increasing the survival of their offspring. The parents are not the only individuals

that hear the calls. Other birds of the same species hear them but do not respond because they have nothing to gain by feeding someone else's young. Predators may also hear the call and use that information to find the nest and eat the young. Some calls are harder to locate than others, and nestling birds give calls that provide few locating clues. Their parents already know where they are and do not need the calls to locate them. When the young leave the nest and move around, however, the parents do not always know where they are. At that time the begging notes of small birds change to a form that is easy to locate (fig. 9.3).

This change is logical, but it is less evident why nestling birds beg at all. It is clearly to the advantage of parents to feed their young. Why should they need begging notes to induce them to bring food? Do the begging notes really stimulate the parents to bring food or do these notes rather represent competition among the nestlings over who will receive the food? Food is a nonsharable resource. That which one nestling gets, another cannot have. Parent birds do not distribute food evenly among nestlings and the vigor of calling may influence who gets fed. Our first hypothesis about why

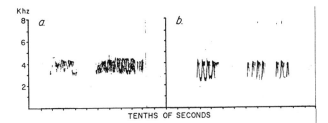

9.3 *The nestling begging call of the yellow-headed blackbird (a) is more difficult to locate than the more strident call given by fledglings (b) after they have left the nest.*

nestlings beg may not be the correct one.

Signaling is also complex because it is not always the best policy to tell the truth. Earlier investigators assumed that natural selection always favored clarifying the meaning of signals so that receivers could predict more accurately what the signaler was about to do. It is somewhat ironic that human beings should have developed this view of communication, because if there is any species in which communication is widely used to deceive and tell half truths it is ours. Any time one politician says something, the first thing another politician asks is not "Is it true?" but, rather, "Why did he or she say that?" We evaluate signals by asking what the communicator hopes to gain and attempt to respond so as to improve our own situation.

The rule governing the evolution of communication signals is, "Does the signal help the signaler?" The advantage sometimes results from being honest, but sometimes it does not. I do not use the terms "honesty" and "deceit" to imply any conscious process on the part of a bird. Blackbirds may be genuinely deceitful, knowing they are telling lies, but we cannot tell if they are or are not. We can observe that what is being communicated seems at variance with reality, or that the signaler is giving only part of the information that might be useful to those receiving it. A viceroy butterfly mimicking a monarch butterfly signals that it is distasteful when it is not. A small male bluegill sunfish that looks like a female and fertilizes eggs by sneaking into the territory of a large male signals that it is a female when it is actually a male.

Signaling occurs in many different situations and for many different purposes. The forms of signals depend on the types of interactions that occur between individuals, how similar or different their interests are, and how important the interaction is. Different signals are used when fighting over food than when competing for a mate. Signals directed toward members of one's own species differ from those directed toward members of other species. Animals and plants signal to predators, pollinators (fig. 9.4), seed dispersers, cleaners and parasite removers, and foraging associates. Intraspecific communication is usually directed to one of three purposes: (a) to improve access to resources; (b) to improve opportunities to mate; or (c) to reduce the probability of predation.

Communication to Improve or Maintain Access to Resources

Interactions over resources are guided by the fact that fighting for them can be dangerous. An individual should not start a fight that it will lose, but the probability of winning may be difficult to judge. Fights among conspecific birds are usually fights among near equals. In equally matched fights, minor and unpredictable events may determine who wins. Even with the best of information there is likely to be considerable uncertainty about the outcome of an encounter.

These conditions have led evolutionarily to behavior patterns that enable contestants to assess, but also to have trouble assessing, one another without actually fighting. Competing individuals often signal their fighting abilities. The winner is usually the individual able to convince its opponent that starting a fight is not wise. Display may involve exaggerating the size of those body parts used directly in fighting. At the same time, individuals should develop good abilities to detect displays that are not backed up by real fighting abilities and should not be put off by mere show.

Encounters differ in terms of what is at stake and the amount of time available to make assessments. Therefore, the elaborateness of competitive displays varies with the situation. Among the most elaborate displays are those associated with territorial defense. Possession of a territory is an absolute requirement for successful reproduction in many species of birds and, given the fact that many individuals live long enough to breed only once, the costs of not getting a territory the first year are great.

Suppose a blackbird arrives in a breeding area and finds a good-quality territory that is already occupied. The resident is signaling his dominance and the newcomer knows that he will be attacked and will have to fight if he attempts a take-over. The resident is likely to conceal whatever weaknesses he has. Moreover, the territory has greater value to him than to the newcomer because he already knows something about it and the resources it provides, and he has already settled boundaries with neighbors. Therefore, he will fight very hard to retain it, perhaps harder than the challenger can afford to fight, given that the territory is, at that time, worth less to the challenger. There is no easy way for a challenger to assess the true fighting ability of a territory holder. If the breeding season is short, there may not be much time to find out.

It has long been known that territory holders nearly

9.4 *A pair of Scott's orioles visits the flowering stalk of an agave.*

always win encounters even if they have held the territory only a very short time. Why this is true is not well understood. Coloration does seem to have something to do with it because male redwings whose epaulets have been blackened have trouble holding their territories. F. W. Peek found that male redwings with blackened epaulets were forced off their territories by new males not previously seen to hold territories in the vicinity, but only if they were blackened early in the breeding season. Douglas Smith found that males were replaced with equal frequency both early and late. In still another experiment, Lynn Morris found that most of her blackened males held their territories, but they did experience greater trespass rates than they had before they were blackened and they did have to work harder to hold their space. They also had more difficulty in forming bonds with females. They engaged in a higher frequency of chasing them than did the control birds who were also captured but had their epaulets painted with a clear material.

These variable results could actually be caused by the same processes. Loss of the red may increase the cost to the male of holding his territory, but whether or not he is replaced may depend on the strength of the opposition and the condition of the experimental male. If he has been stressed by being captured, he may be evicted more readily than if he is in good condition even though blackened. More detailed experiments in which the weights of males are carefully monitored both before and after blackening are needed to detect these subtle but potentially important influences on the ability of males to hold space.

There are even more interesting surprises. Dorothy Mammen has found that territorial male redwings that are captured and held for several days do not usually regain their territories when they are released. Nonetheless, they are present in the area: when she then removed the males who had taken over the territories in the absence of the owners, most original owners quickly reoccupied their former territories. They were actually in the area but were unwilling or unable to regain their territories. She has even been able to remove the territorial male a second time and found that the male who took over the territory the first time did so again.

These results raise a number of interesting questions. Why doesn't the first owner, who successfully repelled all challengers prior to his capture, immediately take over his territory when released? We know that he is there and watching. Why does ability to hold space change so rapidly once the space is occupied? What is the nature of the communication between owner and challenger that influences these outcomes? There is much to be learned about even such familiar behavior as territory defense by redwings.

Territorial displays are among the most conspicuous behaviors of many blackbirds. These involve erection of feathers on many parts of the body, often emphasizing brightly colored patches and body movement, particularly bowing forward (fig. 9.5). Many are accompanied by loud vocalizations, especially songs. The entire character of freshwater marshes in much of North America during the spring is transformed by the territorial displays of male red-winged and yellow-headed blackbirds, perched on the tops of the cattails and bulrushes. The displays are especially obvious because of the high densities of the birds and the openness of the vegetation. The displays of orioles are less obvious because they take place in denser vegetation, but they are equally spectacular, especially because of the bright patches of colors on the bodies of most orioles.

Territorial displays have another interesting feature —they tend to be graded in character (fig.9.6). At first each individual exerts a small amount of effort and escalates its display only if the less intensive version fails to deter its opponent. Fighting is likely to break out only after displayers have escalated to the highest point.

Not all encounters over resources are of this nature. Nonterritorial birds often interact over food items. There is still a winner and a loser, but the stakes are trivial compared with those involved in a territorial contest. Fights over individual food items are handled much more simply. The usual posture is a simple, head-forward display in which few feathers are erected, but the shoulders of the wings may be elevated in preparation for a subsequent attack if the contestant does not retreat. The simplicity of these displays is not owing entirely to the low stakes. Usually the contesting individuals know one another from previous interactions and know their relative fighting abilities. Nothing more than a little reminder is needed. It is when the flocks first form and the individuals are unknown to one another that fighting over food is most intense.

Noncompetitive interactions also occur over food. Adults feed young and females may beg from males and be fed by them, particularly during courtship and while the female is incubating. Begging behavior allows females to evaluate the foraging abilities of potential mates. By feeding their mates males permit them to spend more time incubating than would be possible if the female had to find all of her own food.

Communication to Improve Opportunities to Reproduce

Like defending territories, improving one's reproductive opportunities is an activity with a potentially

9.5 *Territorial or courtship displays of the greater Antillean grackle (top), yellow-headed blackbird (left), giant cowbird (bottom); facing page: olive oropendola (top), Bolivian blackbird (left), and scarlet-headed blackbird (right).*

9.6 The song spread of the male yellow-winged blackbird, like that of many other species, ranges from a form in which almost no feathers are erected to a form in which most feathers are erected and the wings and tail are spread.

9.7 *The precopulatory display of the male boat-tailed grackle, like that of most polygynous species, is vigorous and prolonged.*

high payoff, especially for males who can inseminate large numbers of females. Male birds are normally unable to force unwilling females to accept them as mating partners. The females must be persuaded. Before they can be persuaded they must be found.

No individual can evaluate a potential spouse that it has never previously met. For many organisms finding one another may be the most difficult problem of all, and once a male and a female have gotten together neither is inclined to be very choosy. In fact, among species such as slugs and worms in which encounters occur very infrequently, each individual is both a male and a female and every encounter is an appropriate one for mating.

Birds are highly mobile and have good vision and hearing. Individuals of even rare species can find one another relatively easily, even in dense vegetation. Nonetheless, at the beginning of the breeding season many birds engage in behavior that advertises their presence over long distances. Songs are, of course, the most prominent component of this advertising, but they are also often uttered from very conspicuous perches from which the singer is readily visible over long distances.

Once individuals are in contact, more subtle discriminations are made. Is this individual a good potential mate? How likely is it that a better mate can be found with more searching? Will the present individual already be taken if a longer search is attempted?

Individuals of both sexes gain by presenting themselves in the best possible light because this increases their choices. The more choices one has, the better one is likely to do. Courtship is a process in which each individual attempts to convince its potential partner of its high quality, yet, at the same time, not being ready immediately to commit itself even if the partner shows signs of yielding. For males of species with internal fertilization, there is yet another problem. If the female is already inseminated, the male may be tricked into helping to rear offspring not his own. This risk can be greatly reduced by an extended period of courtship during which the female is closely followed.

Courtship displays of males tend to be conspicuous and are, in general, less graded in intensity than are territorial displays. Even if the same displays are used for both purposes, the most vigorous forms are used when females arrive on the territories. Responses of varied intensity are given primarily to other males. Precopulatory displays are highly variable among blackbirds. Their conspicuousness and vigor may depend on whether they occur before or after pair bonds have been formed. Among the promiscuous colonial grackles, precopulatory displays, which are really courtship displays, are given when females are choosing mating partners. Those precopulatory displays are very elaborate (fig. 9.7). In most monogamous species, however, pair bonds have formed well before copulations start and neither partner is assessing the other at the time of copulation. In these species the female simply assumes a posture which enables the male to mount her (fig. 9.8) and the male mounts with little prior display.

Some species are harder to categorize. Red-winged blackbirds, for example, may copulate both with their territorial partners and with other, usually nearby, individuals. Female redwings often test their partners or invite copulations with neighbors by assuming precopulatory postures near the territory boundary. The males interpret that behavior appropriately because they often attack their neighbors rather than approaching to copulate. They probably detect slight differences in the displays that are not obvious to us but that reveal the intentions of the female. Males, for their part, are constantly seeking opportunities to copulate with other females. The precopulatory display of a male redwing is the most vigorous display in its repertoire. Among yellowheads, where there is much less off-territory sexual activity, males often mount with little prior display.

Communication to Reduce Predation

By signaling the presence of a predator an individual can reduce the chances that one of its associates will be captured. The same signal also tells the predator where the caller is. The caller's survival might be enhanced by remaining silent. Whether or not a call is given depends on the magnitude of risks to the caller and

9.8 Copulation in the monogamous scarlet-headed blackbird is preceded by little display on the part of either bird. The female simply assumes the copulatory posture and the male mounts her.

benefits to the receivers of the signal, together with the degree of genetic relatedness between them.

The simplest communication about predators occurs between parents and their offspring. Parents have a direct genetic interest in the survival of their offspring. Since each parent shares half of its genes with its offspring, it should be willing to take substantial risks, provided an offspring receives at least twice the benefit, or provided that the combined benefits to all offspring are greater than twice the risk to the parent.

Parents do, of course, take great risks in the defense of their young. Blackbirds attack large predators, even making physical contact with them when they come near their breeding areas. It is a familiar sight around marshes in the spring to see male redwings attacking passing crows and hawks, coming in from behind to hit them on the back and neck. The scarlet-headed blackbird in Argentina sometimes lands on the back of the chimango, a scavenging hawk that robs nests. These attacks probably reduce the effectiveness with which the predators can hunt in the area and may induce them to move elsewhere. How much risk the blackbirds expose themselves to is difficult to tell. There are no reports in the literature of blackbirds being captured by a predator they were mobbing. They reduce the risk by always attacking from behind. I am able to deter unusually feisty redwings by turning around and looking at them. They are fully aware of whether the predator can see them as they attack.

Warning flock mates, who may not be related, about the presence of predators is a more complicated matter. The predator alarm calls of most small birds, especially their hawk alarm calls, are difficult to locate. This suggests that the caller is at risk and that selection has favored reducing that risk while still warning others. However, those calls do not tell *either* where the caller is *or* where the predator is. The best response of another individual hearing a hawk alarm call is probably to head for the nearest cover, wherever that is. Conceivably, the individual might flee directly toward the predator. If it had received more information about the location of the predator its response could be better directed.

It has been suggested that hawk alarm calls may actually function to manipulate flock members to the benefit of the caller. Suppose, for example, an individual sighting a hawk simply escapes silently. Its movement might advertise its whereabouts to the hawk. On the other hand, if it remains motionless and gives a call that is hard to locate, it may provoke escape movements by its flock mates, thereby directing attention away from itself. If the caller is really risking itself to help others, calling should be much more frequent in flocks composed of closely related individuals than it is among flocks of distantly related

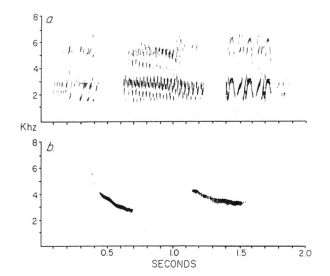

9.9 *The hawk alarm call of (a) the male yellowhead and (b) the male redwing*

individuals. Some birds form breeding-season flocks consisting mostly of pairs and their young while other species form winter flocks that consist mostly of unrelated individuals, but the frequency of use of hawk alarm calls has not been measured in these two cases.

The alarm calls of many blackbirds are easy to locate. Blackbirds of open country are highly conspicuous, and approaching predators probably know where most of the individuals are. In marshes, blackbirds can avoid avian predators simply by diving into the vegetation. The shrill whistle of a male redwing when he spots a hawk is actually rather easy to locate. Yellowheads give an easily located loud, harsh, raucous call when they spot a hawk and they often circle around it while calling. Avoiding detection is not their concern (fig. 9.9).

Prey may also communicate directly with their predators. Some prey signal to the predator about their quality as food. Toxic prey signal their toxicity and predators quickly learn to avoid them; palatable prey mimic toxic ones. Some prey mimic inedible items such as twigs or leaves. Such mimicry is especially common among insects which are small enough to be able to mimic many different objects in their environments. Birds are too large to do that. They depend more on a general matching of their backgrounds.

Prey may also communicate to predators that it is dangerous to attack them. An attack can be risky to a predator because the prey may counterattack. In fact, Geoffrey Parker and F. R. Baker concluded, from an intensive study of the plumage patterns of the birds of Europe and adjacent regions, that signaling to predators of the dangers of attacking them was the most important reason for the evolution of the bright colors of those birds. In support of this idea some of the larger

9.10 *The heads and bills of the large oropendolas and caciques are brilliantly colored. From top to bottom: the black oropendola, the Montezuma oropendola, the olive oropendola, and the Mexican cacique.*

9.11 A female yellow-winged blackbird with her tail up, signaling mild alarm.

blackbirds, those most likely to be dangerous for a hawk to attack, have very brightly colored bills that contrast markedly with their heads (fig. 9.10). Such bills could also communicate danger to would-be competitors as well as to predators.

Increasing the apparent size of the prey is another form of communication to predators. Many insects flash large eye spots when a predator comes very close, thus appearing more like a large animal rather than the small one they really are. Birds do not have such spots, but many of them raise their wings, spread their tails, and erect their body feathers when defending themselves against predators.

Some prey communicate with their predators about their chances of success in a possible attack. This is a common form of behavior among birds and usually takes the form of movements and calls indicating that the predator has been seen. As most adults are very difficult to capture and predators rely on stealth for their success, simply signaling to a predator that it has been seen may be enough to discourage it from attacking. Many sandpipers give conspicuous displays by raising their wings as predators approach. These sandpipers are dull but have bright patterns on the undersides of their wings and on their lower backs, rumps, and tails that are revealed by these displays. Visual signals to predators also occur among blackbirds. Most of the mild alarm calls of blackbirds are accompanied by rapid flicking of the tail (fig. 9.11). In some species the wings are also flicked. A female redwing, for example, repeatedly flicks her wings when a predator is near her nest, although she does so especially when the territorial male is nearby, suggesting that the display is also directed toward him.

10. Black Is Conspicuous and Sometimes Warm

Compared with other mammals, we are highly visual. Our household dogs live in a world of chemical signals hidden from us. Birds are visually oriented, and they also excel in the second sensory mode of great importance to us—hearing. Colorful, singing birds are a major source of pleasure to us. Americans annually spend millions of dollars to go bird watching and to attract birds to their homes and gardens; the selling of bird seed is a major enterprise in the United States.

Most of the bright colors of birds are found on their feathers. Irridescent colors are produced by the structure of the feathers and the way light is diffracted as it strikes the feather and is reflected back toward an observer. These colors change markedly as the light, or as the position of the bird relative to the observer, alters. Most colors are the result of pigments deposited in the feathers, and their appearance does not change much with environmental conditions. The yellow head of the male yellow-headed blackbird looks about the same in all weather and from all angles.

Some birds have brightly colored eyes, bills, legs, and feet. In a number of species these are the most conspicuous parts of the bird. The most striking feature of a male Brewer's blackbird is his bright yellow eye. In the dark understory of tropical forests, the white eyes

and ivory-colored bills of the all-black caciques appear luminescent. Normally one's first sight of the bird is a glimpse of the piercing eye glowing in some dark tangle. In contrast, the feet of blackbirds are very conservative. All of them are black, dark gray, or brown.

Feathers have other functions besides being bright or dull and, during the normal course of events, they wear out. All birds completely replace their feathers at least once each year, usually at the end of the breeding season. In temperate areas, birds look the shabbiest in late summer when old, worn feathers are mixed with new ones, many still growing out and not yet of full size. Ducks and geese molt all their flight feathers at once and cannot fly for several weeks, but most birds

10.1 (Facing page) Plumage patterns of adult blackbirds. The breeding plumage is shown in the left column, the nonbreeding plumage in the right column. For each species the male is above the female. From top to bottom: the spot-breasted oriole (monomorphism), Brewer's blackbird (sexual dimorphism), bobolink (incomplete sexual and seasonal dimorphism), rusty blackbird (complete sexual and seasonal dimorphism), and (below) orchard oriole (complete sexual, incomplete seasonal dimorphism).

replace their feathers more gradually and at no time are they flightless.

Many birds also have another molt, which may be complete or confined to specific parts of the body, that occurs just before the breeding season. This molt increases the brilliance of the plumage and is concentrated in those areas, especially the head, where bright color patches are borne.

Disregarding the plumages of nestlings and fledglings, which may be quite different from those of adults of either sex, there are seven possible combinations of male and female plumages in the breeding and nonbreeding seasons. In monomorphism, males and females are identical and have the same plumage all year. In sexual dimorphism, males and females do not change plumages during the year but the two sexes are always distinct from one another. In seasonal dimorphism, the two sexes always appear alike but they have different breeding and nonbreeding plumages. In incomplete sexual and seasonal dimorphism, females retain the same plumage all year but the males change from a nonbreeding plumage like that of the female to a distinctive breeding plumage. In incomplete sexual but complete seasonal dimorphism, both males and females change to a distinct breeding plumage. The sexes are alike during the nonbreeding season but different in the breeding season. In complete sexual but incomplete seasonal dimorphism, the female has the same plumage all year. The male changes plumage but always appears different from the female. Finally, in complete sexual and seasonal dimorphism, individuals of both sexes change plumages but always appear different from one another. Of these seven possible combinations, six are found in one or more species of blackbirds. Only the fifth, incomplete sexual but complete seasonal dimorphism, is absent. Examples of some of these patterns are shown in figure 10.1. As a group, blackbirds are highly variable in plumage patterns. In one-third of the families of North American birds, every species is completely monomorphic. It is among the blackbirds and the families of birds most closely related to them, the tanagers, warblers, and finches, that great variety in plumage systems is found.

We do not know how color patterns have evolved in any particular species, but it is clear that many different environmental factors have been involved. Among them are energy conservation and control of temperature, concealment, and communication with other individuals of the same species or with individuals of other species, such as competitors and predators. It will be useful to have a general idea how each of these factors influence coloration before we examine the color patterns of blackbirds.

Color and Energy Conservation or Control

For a given size, a black bird absorbs the most and a white one the least amount of solar radiation. If staying warm is an important need for a bird, then black should be the best color. If overheating is the most important problem, then whiteness should be ideal. However, temperature regulation problems seldom stay the same. In most environments it is beneficial to take up heat from the sun at some times of the day or year, but to avoid taking it up at other times. The best solution is to change colors quickly in response to environmental conditions. Some animals, such as lizards and fish, can do this, but the color of a feather is fixed when it grows out; the bird is stuck with it until the next molt. At best, birds can have winter and summer plumages or breeding and nonbreeding plumages.

Birds maintain relatively constant and high body temperatures. Their temperatures are even higher than ours—about 108 degrees Fahrenheit (42 degrees Celsius). Also, like ours, their normal operating temperatures are very close to the upper lethal limits. Death in people and birds occurs when body temperature is raised only a few degrees. In most environments most of the time, the body temperature of a bird is above that of its environment, but birds still must often unload heat. An active bird generates more than enough heat to maintain a high body temperature even when the outside temperature is quite cold. The thicker the insulating layer provided by the feathers, the more difficult it becomes to unload excess heat and the greater the problem of being in the sun.

One hypothesis to explain the existence of the many all black and all white birds is based on where they sleep at night. At night a sleeping bird generates less heat than at any other time, but it is exposed to the coldest temperatures. The thickness of feathers may be determined primarily by the amount of insulation a bird needs to maintain a high body temperature on cold nights. This depends on how cold it is in general and the micro-temperatures an individual finds or creates in its sleeping places. The coldest sleeping places are in the open where a bird loses heat by radiation to the nocturnal sky which, on clear nights, has a temperature of -273 degrees Celsius. Heat loss under such conditions is extreme. Birds that sleep in the open need thicker insulation than those that sleep in sheltered sites.

Birds that sleep in the open may be white because they need so much nocturnal insulation that they have problems with excess heat during the day. Birds that sleep in sheltered places do not lose as much heat at night. They need less insulation and are not faced with such severe overheating problems during the day. This hypothesis is supported by the fact that most all-white

10.2 Common grackles huddle together under the protecting branches of a pine tree to save energy on a cold winter's night.

birds sleep in the open (herons, gulls, terns, snowy owls), while all-black ones, which do have a lower weight of feathers per unit of body weight than do white ones, tend to sleep in sheltered sites (fig. 10.2).

So far so good, but this does not explain the existence of many all-white and all-black tropical birds. These species never experience even cool temperatures, and absorbing solar energy during the tropical day seems undesirable. Nonetheless birds do benefit from intake of solar energy even in very hot climates. Part of the reason lies in the fact that the thermal properties of black plumage change with wind velocity. Solar energy penetrates deeply into the feathers of a white bird but is mostly absorbed superficially by black feathers. If it is windy, a white bird may actually absorb more useful solar radiation than a black one, while the reverse is true when it is calm. A black bird in a hot climate may actually reduce the energy it expends to maintain a constant body temperature by taking on solar heat at dawn when it is relatively calm, but still have a low heat load during the middle, hotter, but windier part of the day.

Temperature control and regulation is not the only factor influencing color. If it were, all birds would probably be either black or white or some combination of the two. Other colors have poorer thermal properties, but they have advantages for dealing with predators, prey, and associates.

Color and Concealment

It is a familiar fact that many animals are hard to see against their normal backgrounds. Their colors and patterns blend in so well that we may actually touch them before we realize that they are there. A well-known concealing pattern is *countershading*, where the dorsal surface of an animal is darker than the ventral surface. Because sunlight comes from above, the back of a bird normally casts a shadow on its belly. If a bird were uniformly colored, its belly would appear

10.3 *A meadowlark, with its streaked brown back, is very inconspicuous when it crouches in open grassland.*

conspicuously darker than its back. By being counter-shaded, it appears uniformly colored and more difficult to see. Many open-country birds are countershaded, but this pattern is rather rare among icterids, being confined to some grassland and marsh-nesting species (fig. 10.3).

If conspicuousness is favored, then a reverse pattern in which the dorsal surface is brighter than the ventral is an excellent one. Among blackbirds there is one striking example of such a pattern, the breeding plumage of the male bobolink (fig.10.4). With their black ventral surfaces and bright white and buff nape and back patches, male bobolinks are very conspicuous when perched on the tops of stalks of herbaceous vegetation in the grasslands where they breed or when in flight over their territories. And as soon as they are done breeding, they molt back into the cryptic countershaded coloration that they have most of the rest of the year.

Background matching is the basis for most cryptic coloration. Countershading does not work if the colors are wrong. Backgrounds are seldom of a single color and, as a bird moves around, it encounters differently colored environments. In part this can be handled behaviorally. The individual can avoid those areas where its own plumage matches poorly while seeking out those areas where it matches better. This, in turn, raises other problems, because good foraging areas are not necessarily the best areas for background matching. A hungry bird is likely to forage in more dangerous places.

Whether or not a pattern matches its background depends on the distance from which it is viewed. The color pattern of a streaked, mottled bird, such as a female red-winged blackbird, is very conspicuous to

an observer a few feet away, but it blends well into the background when viewed from a distance. This simple fact provides a clue to much of the concealing coloration of birds. Danger comes generally from predators searching from a distance. At this scale of perception, mottling and streaking, as long as the overall sizes of these color patches are about the same as the streaking. of the plants and soils among which the bird is found, offer excellent concealment. This is so even if many different colors, some of which may not actually be present in the real background, are found in the pattern. Some of these patterns work quite well against a variety of backgrounds provided overall color intensity does not vary greatly.

Viewed from a distance a female redwing appears to be a very dull, streaked, and drab bird. She is an uninspiring sight compared with her brilliant and conspicuous male. Nonetheless, closer examination reveals a rich array of dorsal and ventral color patterns including whites, blacks, reds, oranges, and various shades of brown. Each female is quite distinctive. There is an almost infinite variety of patterns with different emphases on the amount and boldness of patterning on the head, throat, breast, and back. The gaudy males, on the other hand, are all alike. Redwing female coloration is actually much more interesting than male coloration.

Color and Social Signals

Birds use their feathers, particularly those that are brightly colored, in displays. Coloration may not have originally evolved for signaling purposes but it is certainly used for that purpose now. Coloration can potentially show a number of things. It reveals the species to which an individual belongs, except in a very few cases where we (and presumably they) must rely on vocalizations. Coloration may also signal the sex of an individual although there are more birds in which males and females have identical plumages than species in which they are different.

Color can also signal the status of an individual. Harris's sparrows with large amounts of black on their heads and breasts are high in the winter dominance ranks while those with little black are low in the peck order. The amount of black is actually correlated with fighting ability. Bleached birds are attacked by birds that had been lower in rank, but the bleached individuals are successful in defeating those challengers. In fact, they become very aggressive because they are attacked so often by birds "who should know better."

Color can also signal probable future behavior, and it can reveal the size and shape of its bearer. A bright rump patch that is normally concealed by the wings signals flight when it is suddenly exposed. The same is true for colors under the wings or at the margins of the tail which are usually concealed by other features.

The Colors and Color Patterns of Blackbirds

As their name implies, black is a dominant color among blackbirds, but there are blackbirds with no black in their plumage and few blackbirds are entirely black. Certain colors are widespread in the family while others are rare or lacking. The most common colors are

10.4 *The male bobolink, whose dorsal surface is brighter than the ventral, illustrates the most conspicuous plumage patterns possible. The female, with darker back and lighter belly, is very inconspicuous in the grass.*

black, brown, chestnut, red, orange, yellow, olive, gray, and buff. White is mostly confined to small patches, being present in the wing bars of a number of orioles, in the back and rump of male bobolinks, as a strip below the epaulet in male tricolored blackbirds, and as part of the background color in some of the mottled female blackbirds such as redwings, tricolors, and yellow-wings. Male yellowheads have small white patches in their wings which are exposed when the wings are spread. Undertail coverts of many orioles and females of other species are also white.

Colors that are entirely missing in the family include violet and green (other than olive-green). Blue is absent as a feather color among icterids, although it is present on bare skin around the eye in a few oropendolas and in the colorful troupial. Blue, green, and violet are common feather colors in related families, such as warblers, tanagers, and finches. Their absence among blackbirds may be a historical accident, but since colors leave no fossil traces we do not know.

The patterns that characterize different species are not maintained because there has been a lack of plumage variability upon which natural selection could act. Measurements of a population of redwings in Ohio revealed that 5 percent of adult males had a concealed band of white feathers across their breast. A full 86 percent of adult males and 88 percent of immature males had considerable white on their tail feathers. A territorial male redwing in Madison, Wisconsin, had a buffy throat. Male yellowheads differ strikingly in the amount and intensity of yellow they have. Despite the occurrence of variability, natural selection usually maintains characteristic species-specific plumages. Deviants must not survive and reproduce as well as normal individuals.

Cryptically colored blackbirds with a predominance of streaked brown patterns in their plumage, especially dorsally, are found in grassland habitats and among females of marsh-nesting species. Female bobolinks, and males outside the breeding season, are obscure brownish birds that are difficult to see in the tall, grassy places where they live. Meadowlarks of both sexes combine streaked back plumages with bright ventral coloration, yellow in the northern hemisphere species and red in the southern hemisphere ones. A crouching meadowlark conceals its bright colors, presenting only the cryptically colored back to the view of predators.

The distribution of colors is different among marsh-nesters. Males of these species are basically black, usually with bright color patches. The females reveal the species to which they belong by having some brighter color in the places where the males are most vivid, but their aspect is dominated by streaked, cryptic patterns.

Dark, uniform coloration is found in two major groups of icterids. One forages solitarily or in small groups in the dark understories or lower canopies of dense tropical forests. The solitary caciques all have conspicuous whitish bills but otherwise are completely black. The velvet-fronted blackbird of the upper Amazon River Basin is also completely black. Unlike the caciques it has a short, dark bill, and it forages in small groups in trees along rivers and streams. It is usually in brighter light than the understory caciques but not in such bright light as the canopy species. Still higher up in the forest are basically black species with patches of color that can be concealed during normal activities, but displayed when desired. Among these species are the canopy-inhabiting caciques with bright rumps and wings and the virtually unknown species of the genus *Macroagelaius*, both of which have tufts of bright feathers under their wings (fig. 10.5). Like the velvet-fronted blackbird they have short, dark bills.

The other major group of all-black icterids forages terrestrially in open country. Included in this group are all cowbirds except the baywing, grackles, the chopi, and the melodious, austral, Bolivian, Brewer's, and rusty blackbirds. There are also two South American species of *Agelaius*, the unicolored blackbird and the pale-eyed blackbird, breeders in marshes or rank herbaceous vegetation, in which males are all black. Females in the species with all black males are also unicolored, but they range from being about as black as the males, as in the melodious blackbird and chopi, to being noticeably browner and, sometimes, streaked.

Males of another group of blackbirds are black with small to medium patches of color, usually on the wings, head, and rump. The familiar marsh-nesting species of temperate North and South America fall into this group, as do some tropical caciques. Some of the color patches, particularly those on the wings, can be concealed, but others, such as those on the head, cannot. Browns are prominent in the plumages of females in many species in which males are primarily or entirely black.

The base color of male orioles is black, but the amount of bright coloration is usually greater than in other icterids. Orioles range from those that are all-black except for epaulet colors to species in which oranges and yellows are the dominant body colors, the black being confined primarily to the wings and tail (fig. 10.6). Among tropical orioles sexes are identical, while in the migratory species females are much duller than males.

The basic body coloration is brown, chestnut, or olive rather than black in most oropendolas. Only the black oropendola has a primarily black body. All oropendolas

10.5 *The golden-tufted grackle (Macroagelaius imthurni) below, and the mountain grackle (M. subalaris) above, both have long, yellow or orange feathers that are visible only when the birds raise their wings.*

10.6 Plumage patterns among orioles. The brightly colored orioles fall into two groups: those for which the entire head is black and those for which only the throat and upper breast are black. Orioles also vary in the amount of bright coloration on the back, wings, and tail. Pictured here, from top to bottom, are the white-edged oriole, spot-breasted oriole, streak-backed oriole, bar-winged oriole, and black-cowled oriole.

have bright yellow outer tail feathers from which they acquired their colloquial name "yellowtail" in parts of South America.

Sexes of the bay-winged cowbird are identical, brown with rusty patches in their wings, a plumage quite unlike that of other members of the genus. It is not clear why such a different color has evolved in baywings, but they do show that if a different color is advantageous, it evolves even in genera that are otherwise very similar in color.

The yellow-rumped marshbird and brown-and-yellow marshbird, inhabitants of wet grassy places in Argentina, Uruguay, Brazil, and Paraguay, are quite unlike other icterids. Their olive-brown body is offset by bright yellow patches on the shoulders, under the wings, and on the belly. Sexes are identical.

Why Blackbirds Have the Colors They Do

The fact that all-black blackbirds occur in two very different environments suggests that different factors favor blackness in the groups. Blackness probably provides inconspicuousness for caciques of the forest understory. Their blackness can have little to do with capture of solar energy because they rarely encounter sunlight. Bright colors would not show up well in those situations even if they were available to be used. Most birds of forest understories are dull-colored.

Blackness in open-country birds is another matter. Blackbirds in grasslands and marshes are highly conspicuous. In fact, black is probably the most conspicuous color in those environments. These black species travel in flocks much of the year. Their defense is based on their ability to detect predators at great distances, far enough away to avoid being taken in a surprise attack, and on their being very fast fliers that cannot be captured in open pursuit.

When I first began my blackbird studies I watched redwings in a small marsh in the hills above Berkeley, California. The marsh, in a small valley, was surrounded by many oak and eucalyptus trees. Conditions were highly favorable for hawks to launch surprise attacks. I witnessed several such attacks by Cooper's and sharp-shinned hawks, but none of them were successful. An adult redwing can easily pull away from a Cooper's hawk in level pursuit. The birds were alert enough that none were taken by surprise while I watched. My presence may have deterred hawks, but since about 65 percent of adult redwings survive between breeding seasons at the Potholes, they are obviously effective in avoiding predators.

Nonetheless, there are times and places where it is better to be more difficult to see. The males of a number of North American black blackbirds molt into a rather different plumage at the end of the breeding season. The new feathers of redwings are broadly tipped with brown and buff, especially on the back, the part most often seen by predators. The tips of the new feathers are so brightly colored in the rusty blackbird that it has been named because of its winter plumage (see fig. 10.1). In the spring, male rusties are glossy black with yellow eyes. Male redwings and rusties acquire their black spring plumages through wear of the feathers so that by late winter only small traces of the lighter tips remain. These males would not acquire this mottled plumage if there were not some advantage to being less conspicuous during the fall and winter.

The cryptic coloration of the females of some species also suggests that there is some advantage to being concealed. Since female blackbirds build the nests and incubate the eggs by themselves, perhaps that is the time when being hard to see is most important, but there are many species in which the females are bright—some even sing from their nests while incubating.

The value of being cryptic is further suggested by the plumages of the more solitary grassland species, especially meadowlarks. Meadowlarks are less powerful fliers than the flocking blackbirds, and may be especially vulnerable to avian predators in the sparse, open, grasslands where they live. Their major defense is to crouch on the ground, concealing their brilliant yellow or red breasts and bellies. Blackness may be a luxury that solitary grassland species cannot afford because risks of predation are too high. Only if easy escape is possible and if many eyes are watching can the advantages of blackness, whether they be thermo-regulatory or communicatory, be afforded. This may be why females have become black in the species derived from redwings in Cuba and Puerto Rico, where there are few bird-eating hawks.

The position of the bright colors on meadowlarks suggests that there may be selection both for inconspicuous plumages to avoid predation and for bright signaling colors. Other color patterns also suggest a compromise between these two powerful selective forces. Many color patches are located where they can be displayed or concealed at will (fig. 10.7). This is true of rump patches, shoulder patches, underwing patches, colored outer tail feathers, and breast and belly patches. A redwing male can sit with his epaulets totally concealed. He can also sit with just the yellow strip below the red showing or with the entire epaulet visible and erected. White outer tail feathers can be revealed by spreading the tail. Rumps can be concealed by folding the wings over the back or revealed by letting the wings droop a bit.

Not all color patches of blackbirds are in locations where they can be concealed. A yellowhead can do little to hide its bright head. The same is true of the

10.7 Bright colors on many blackbirds are located where they can be displayed or concealed when desired. The epaulet oriole (upper left) has yellow shoulders that can be completely covered when the wing is folded. The golden-winged cacique (upper right) has concealable wing and rump patches. The brown-and-yellow marshbird (center) has, in addition to its yellow belly, a large yellow patch on the undersides of its wings. The Baltimore oriole (lower left) has bright patches on its wings, rump, and tail that are much more conspicuous when the bird displays. The red breast and shoulder patches of the lesser red-breasted meadowlark (lower right) can be covered by folding the wings and crouching.

128

10.8 *The males of the race of redwings living together with tricolored blackbirds in California lack the yellow band underneath the red shoulders.*

diminutive yellow-hooded blackbird of northern South America, the oriole blackbird, the many orioles with mostly yellow and orange heads, and the brightly colored oropendolas whose heads may be a veritable rainbow of colors, mostly owing to bright bare patches of skin and multicolored bills.

Color patterns may evolve in response to other species living in the same area. Over most of North America no other blackbird looks very much like a redwing. In lowland California, however, redwings breed in the same marshes with the closely related tricolored blackbird. Tricolors differ from redwings in having a white strip below the red epaulet rather than a yellow one as redwings have. Where the two species live together, however, redwings have lost their yellow strips and have epaulets that are entirely red (fig. 10.8). These birds were once considered a different species— the bicolored redwing. Now we know that they are only a well-marked race of redwings connected to the others through a series of intermediates.

In the southern part of the Mexican Plateau something even more interesting happens. Populations of breeding redwings occur in a roughly doughnut-shaped area, in one part of which breeding populations

of two different races of the redwing overlap. One of those races also has the "bicolored" pattern. The two races formerly occupied different parts of the marshes in the valley of the Río Lerma, but we are not likely to have an opportunity to study that situation more thoroughly because most of the marshes have been drained and converted to agricultural production.

Why Birds Have Evolved Bright Colors

The use of colors in communicative displays suggests that coloration makes it easier to alter the behavior of other individuals. Nonetheless, a great deal of coloration could evolve simply by selection for concealment. The most effective plumage for concealment in a grassland would not be a uniform color. Stripes and streakings help to break up a bird's outline and make it more difficult to see at a distance. Close-up, however, such patterns could be very striking indeed. But this doesn't explain why coloration enhances the effectiveness of signals.

One possibility is that by being more conspicuous, a signaler is more likely to be noticed by others. If changes in behavior occur as a result of witnessing a display, coloration would enhance its effectiveness. Bright colors and erection of feathers make an individual appear larger than it really is, thereby intimidating opponents. The territorial male redwings whose epaulets were blackened by Douglas Smith (see p. 109) had more difficulty in holding their territories than males who had their epaulets painted with a transparent paint. Unfortunately, these results are difficult to interpret. Since the epaulets were totally obliterated, the other males may have mistaken them for birds of another species. Or perhaps the experimental individuals knew that they had lost their epaulets and changed their behavior. Would the results have been different if the epaulets had merely been altered in size? The experiment shows that the epaulets are in some way involved with communication, but it does not tell us how or why.

Two very different points of view of the communicative significance of coloration have emerged in recent years. One holds that colors help exaggerate the size of the displaying individual and tend to dazzle opponents or potential spouses. The other states that colors have evolved to make communication more accurate by enabling other individuals to better assess the size and shape of the displaying bird. Supporters of this view doubt that other individuals are easily fooled by bright colors. In fact, they argue that colors tend to evolve precisely in those places on the body that make it difficult for a misleading signal to be sent. Only colors in those positions will receive attention from others for the very reason that they are reliable. Both points of view are probably partly correct.

We can use the conditions favoring accuracy or bluff in signaling to predict the circumstances in which signals should be more or less reliable. Male birds that defend territories during the breeding season should tend to be more variable in color during the non-breeding season than they are in the spring, because there is only one way to be successful in spring whereas both dominants and subordinates survive during the winter. In species where females also defend territories, the same should be true for both sexes but, in those in which defense of space is confined to males, females should be as variable during the breeding season as outside it.

The plumages of red-winged blackbirds show these patterns nicely. Males during the breeding season are extremely uniform in coloration. There are differences in their sizes and shapes, but the same feathers are colored red or yellow in all individuals. First-year males, who do not usually hold territories, are different from older males, having more brown on their wings, tail, and body feathers and more orange and blotches of black and brown in their epaulets. They are also highly variable. Some look almost like females, others almost

like adult males (fig. 10.9). Female redwings are extremely variable during both breeding and nonbreeding seasons. As with males, some of this difference is related to age because all first-year females have very dull plumages, but after the first year age is not a good predictor of brightness. Some females actually become duller as they get older. And a female that is bright on her back may be dull on her wings, and *vice versa*. The earlier females to breed in the spring are invariably brightly plumaged, and only these females engage in prolonged territorial behavior. Later-breeding females, which may be either bright or dull, engage in little or no defense of space but immediately begin nest building. Brightness of coloration influences interactions among females, but it is not necessary for a female to hold a territory to breed successfully. All females capable of breeding appear to do so. No student of redwings has found a floating population of nonbreeding females. Nonbreeding males are always present, including not only first-year birds but older ones as well.

Territorial behavior may be a powerful force favoring uniformity in plumage, but it cannot be the only reason, because many birds are nonterritorial and yet are nonvariable in plumage. Courtship behavior may also be important in maintaining plumage uniformity among these species. Mate selection, like territory selection, must often be made in a hurry, and getting full information about potential mates is impossible. Even among people, where courtship may be enormously prolonged, couples are usually in for some real surprises when they get to know each other. Courtship is also a situation where there is no benefit in signaling one's weaknesses. Accurate signaling may reduce the chances of obtaining a mate and delay pair formation. Mate choice has all the properties needed to favor vigorous signaling.

Interactions with predators also favor deceitful signaling. Predators must make quick decisions about whether or not to attack a prey individual, and attacking the wrong individual may be dangerous. Prey benefit by signaling that they are in good condition and will be difficult to catch and that they are dangerous.

Why Colors Are Where They Are

Part of the answer relates to the value of being able to conceal colors when they are not being used, but this can explain only part of color patterns. Most of the colors of orioles are in places where they cannot be concealed by assuming some particular posture. Their plumages are variations on two basic patterns. In one, the entire head is black but the amount of black on the back and the extent of wing bars and other colors on the wings is variable. In the other pattern, only the

10.9 *Plumages of first year male redwings vary from femalelike to almost adult malelike.*

throat is black. The rest of the head is primarily yellow or orange. This pattern is also associated with variable back and wing coloration. Rather striking plumage differences apparently have rather simple genetic bases because Baltimore (black-headed) and Bullock's (mostly orange-headed) orioles, despite their striking differences, hybridize in the North American Great Plains.

Why boundaries between color patches occur where they do has never been explained. Some boundaries occur at the ends of body structures but others do not. Patterns are not necessarily determined by where feathers end because individual feathers can and do have very complicated color patterns. Do these boundaries tell us what structures are most important in that species and which ones potential mates or competitors assess most carefully, or are they primarily the result of historical accident? We need many more experiments in which the boundaries of color patterns are altered to help us answer these questions.

Why Plumages Vary with Age and Sex

One of the obvious differences among birds is that in some species males and females are identical in appearance while in others they are quite different. In some species, the young immediately acquire a plumage like that of the adults while in others they first acquire an immature plumage distinct from that of the adults of either sex. The length of time an immature plumage is worn often differs between the sexes. Males commonly retain an immature plumage through the entire first year of their life while females rarely do so. Among blackbirds females in their first breeding season are very similar, if not identical to, older females, while first-year males in a number of species are strikingly different from older males.

Sexual differences in plumage are strongly correlated with the roles played by the sexes, in particular who is responsible for the defense of the territory. The pres-

10.10 First-year males of the northern migratory orioles have plumages very much like those of females except that they have black throats. Shown here are the hooded oriole (top left), Baltimore oriole (lower left), Scott's oriole (upper right), and orchard oriole (lower right). First-year males are shown below older males of each of the species.

ence of immature plumages is also strongly related to the form of the social system. Immature male plumages are prevalent among species with polygynous mating systems in which some of the males dominate access to the females, leaving other males without any breeding opportunity that year. Among colonial species, where males cannot readily prevent others from gaining access to breeding sites, immature male plumages are absent.

Interestingly, most immature males look very much like females. First-year male temperate-zone orioles all look like females of their species except for having little black throat patches (fig. 10.10). First-year male yellowheads look like overgrown females with somewhat brighter yellow on their throats. As already mentioned, first-year male redwings are highly variable in plumage, in part owing to the age of the male when he underwent his first molt in late summer. Earlier-hatched males molt into plumages more like adult males than do later-hatched males. But knowing this correlation does not explain why it should exist. It does not cost more energy to assume the adult plumage than a more female-like plumage, so there must be some other advantage.

Sievert Rohwer tested the responses of territorial redwings to the presence of intruding young males with different plumages. First-year males looking more like adult males were more quickly attacked by the territory holders than were more female-like young males. The latter were, in fact, likely to be courted initially. Males with dull plumages can more easily move in and around the territories of adult males than brighter ones. This may enable them to copulate with females when the teritory holder is absent, but all "extramarital" copulations so far observed among red-

wings—and quite a few have been witnessed—have been between a female and an adjacent territorial adult male, not with a young male. Alternatively, a young male may improve the possibility that he will be able to take over the territory in a future year. By being on the territory and learning about it, he may be better able to use it in the future.

If so, why aren't all first-year males dull? First-year males do sometimes establish territories and attract females. Only the more adultlike of young males have any chance of doing this, and older first-year birds who are a bit more experienced than their younger "peers" may be intrinsically better at this. This would explain the correlation between age of male and the brightness of his first-year plumage.

There are only a few species of birds yet to be discovered. All of those that have been named have been described and measured. We know how the males and females appear and, in most cases, we also know the plumages of the fledglings. Yet coloration remains one of the most poorly understood of avian characteristics.

Color patterns are central to the ways in which birds evaluate and compare one another. It is this comparative role of color that probably produces the amazing stability of the plumages of bird species over space and time. This makes bird-watching much easier but the study of coloration much more difficult. Conventions are much more difficult to untangle than adaptations to the physical environment. They have unique individual histories and may not be correlated with obvious environmental features. Yet, like human dress customs, avian color patterns may evolve as they do because birds recognize plumage deviations as threats.

*11.1 A pair of melodious blackbirds sings a duet to advertise
their territory.*

11. Talk Is Cheap

Blackbirds are noisy! They are especially noisy when they are breeding, but some are vocal all year. Because we are a very talkative species, we tend to notice the vocal animals around us, and the muteness of most of the living world tends to escape our attention. Plants are all voiceless, as are most animals. This makes vocal animals even more interesting (fig. 11.1).

During the summer of 1974, when I was engaged in field work in the deserts of northwest Argentina, I repeatedly heard a strange, loud, yelping call consisting of a series of notes rising in pitch. I strongly suspected that it was produced by a mammal, but neither I nor my associates could think of a local species likely to utter it. The Andalgalá mystery mammal call was a frequent item of inconclusive speculation over cool drinks at the end of a hot day's field work.

One morning while I was seated under a mesquite tree watching a nest of bay-winged cowbirds the mystery call sounded closer than I had ever heard it before. I immediately terminated observations at the nest and stared in the direction from which the call had come. Soon it was repeated, still closer to me. I remained motionless. In a few minutes some sheep appeared, followed by a mangy dog, and another call. Shortly a gaucho on his horse rode into view, uttered the call not more than fifty feet from me, and then disappeared into the mesquite trees down the wash. He was never aware of my presence. The mystery Andalgalá mammal call had been identified.

This experience made very clear to me how vocal signals are valuable to animals. They can be heard when the caller cannot be seen. This was, I later learned, the function of the Andalgalá call. The gauchos call to remain in contact with one another as they herd their sheep across the countryside. The call may also have reminded the sheep that the herder was immediately behind them. Talking has another advantage—it is cheap. One can make a lot of noise for very little effort, a fact well known in human discourse. It is also possible to make a lot of noise without saying very much.

From the scientific point of view, vocal signals have another very important feature: they have evolved purely for communication purposes. An animal can remain silent if it wishes to. It cannot refrain from sending visual signals. Merely by being visible an individual inevitably sends visual information about itself, whether or not there is any intent to signal.

Metabolic activity also generates odors whether we want it to or not. But an individual can do everything silently, as most animals do. Therefore, vocal signals are not such complex compromises between signaling and other functions as are visual and chemical signals.

Not only can an animal give a call while remaining hidden, but the structure of calls can be modified to make them easier or more difficult to locate. Terrestrial vertebrates locate the source of a sound by comparing, at the two ears, three types of events. The first is the time of arrival of a sound. This difference is easy to detect if the sound starts and stops abruptly but difficult to detect if it starts and stops gradually. The second is the difference in loudness at the two ears caused by the sound shadow created by the head. The effectiveness of the shadow depends on the pitch of the sound. Low-pitched sounds tend to bend around objects more easily than do high-pitched ones. Consequently, it is easier for a vertebrate to tell the source of a higher-pitched sound.

Finally, vertebrates compare the phase of the sound wave at the two ears. Because of differences in time of arrival, at any moment the particular point on the wavelength of the sound is different at the two ears. This difference is most useful for lower-pitched sounds and becomes more ambiguous at higher-pitched sounds because many wavelengths may intervene between the two ears.

A sound that consists of a single long note that starts and stops gradually is harder for a vertebrate to locate than a sound with a wide range of pitches and many abrupt starts and stops.

A sonagram plots wavelength frequency against time (fig. 11.2). It shows what different pitches (or, more precisely, wavelengths) are present and with what intensity over the course of a call. Higher-pitched sounds are higher up on the sonagram, lower-pitched ones lower down. Louder parts of the sound show up darker than softer ones. Sonagrams tell us much more about the structure of a sound than can be determined by listening to it. Precise measures can be made of the intensity, duration, and pitch of components of complex sounds. The invention of the sonagraph—the machine on which sonagrams are made—was stimulated by the desire to teach deaf people how to speak by showing them a picture of the sound they were attempting to make so that they could compare it with the picture of their own sound. The value of these

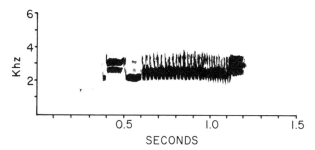

11.2 A typical song of a male red-winged blackbird, showing the introductory syllables, middle syllable, and terminal trill.

sound pictures was quickly recognized by zoologists and their use has revolutionized the study of animal communication.

The Vocal Communication Environment

Having established that vocal signals evolved strictly for the purpose of communication and that they differ in the ease with which they can be located, we are in a position to examine the situations in which animals communicate vocally and how those conditions influence the structure of the sounds they produce.

Animals communicate vocally over highly varying distances. Humans may engage in nose to nose shouting matches or they may, like the Andalgalá gauchos, communicate over very long distances when they are not in visual contact. For a given amount of energy expended in their production, high-pitched sounds carry farther than low-pitched ones. If there are objects in the pathway of a traveling sound, however, low pitches penetrate better. This is why birds that sing in the forest understory, where the sounds must often travel through dense cover of bushes, sing low-pitched songs while those that sing from the tops of the trees have higher-pitched songs.

The structure of vocal signals is also influenced by other sounds being produced in the environment. Some of these sounds are, and some are not, produced by other animals. Wind passing through the leaves and branches of trees produces characteristic sounds. A bird singing in different environments must overcome rather different background noises. Many of the sounds in any environment are produced by other animals, and these sounds interfere with the sending of signals. When birds are recorded in the field it is often difficult to get clear songs from a single individual because so many other birds are singing nearby at the same time. The sounds of *other* species are important primarily because they interfere with transmission and detection. Individuals of the *same* species have special significance because they are potential competitors for resources or mates, or because they are group members with whom

to remain in contact. Therefore, students of animal communication believe that interactions with members of one's own species are a greater force in the fashioning of a species' vocalizations than are the utterances of other species.

To increase the efficiency of transmission of a sound over longer distances, an animal can use a low pitch, concentrate more energy in one frequency rather than scatter the energy over a broad range of pitches, make the sound louder, break it up into shorter bursts, and give it a distinctive rhythm. These were exactly the features of the Andalgalá mystery mammal call. It consisted of a burst of sounds rising in pitch concentrated in a single frequency at any point in the sound. I could hear it miles away and knew fairly accurately where it was coming from even though I couldn't see its producer.

What Sounds Do Blackbirds Utter?

Using these general principles about how the structure of sounds influences their transmission and the ease with which their source can be located, let us look a bit more closely at the structure of blackbird vocalizations. The utterances of blackbirds, like those of other birds, fall into two main classes—songs and call notes. Songs are usually the longest and most complex sounds that birds make. Call notes are often single-syllabled sounds (fig. 11.3), but longer calls can be formed by stringing together a number of call notes of the same or different kinds. There is no absolute dividing line between songs and calls, but the categories are still useful.

Songs are used basically to advertise the presence of the singer. Among territorial species songs communicate possession of a territory. Females hearing the song know that a male with some real estate is interested in establishing bonds for the purpose of reproduction. It is a "have genes, will share" signal. Other males hearing the call know that the singer possesses the territory, which it will defend against all challengers. In both cases, the singer gains by letting other individuals know where it is. Most songs have structures that make them very easy to locate. They are loud and complex with wide frequency ranges and many starts and stops.

The structure of blackbird songs varies with the distance between singing males. Among species that defend large, exclusive territories, even the nearest neighbors are usually far away and potential mates are attracted from great distances. Icterids such as bobolinks, meadowlarks, and orioles that defend large territories sing musical songs in which energies are concentrated on a single pitch at any one moment (which is why they sound musical), in which pitches vary during the song, and which have a lot of starts and stops (fig. 11.4). Many other icterids, however, are

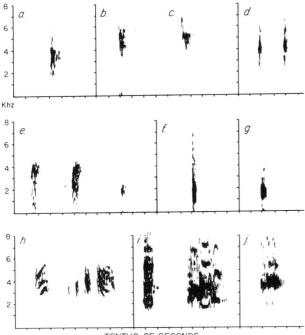

11.3 Call notes of various blackbirds. (a) "check," brown-and-yellow marshbird; (b) "check," yellow-winged blackbird; (c) "chick," yellow-winged blackbird; (d) "check," scarlet-headed blackbird; (e) "check," red-winged blackbird; (f) "check," yellow-headed blackbird; (g) "check," tricolored blackbird; (h) Austral blackbird; (i) yellow-rumped cacique; (j) russet-backed oropendola.

colonial or have small, clumped territories. They sing lower-pitched, nasal, growly songs (fig. 11.5). There is an analogy between the problems of communicating in a dense colony and a problem of psychology known as the cocktail party effect. In a colony, or in a cocktail party, there is a high level of background noise and individual sounds are hard to distinguish. A wide spectrum sound, such as the growly songs of colonial icterids, allows a receiver to make simultaneous comparisons at both ears at many frequencies. These growly songs are probably easier to locate under these conditions than are the more musical songs of territorial icterids.

Call notes are also intended for listeners at varying distances from the caller. Sometimes the caller benefits by not revealing its location. This is especially true for alarm calls given in the presence of dangerous predators (see fig. 9.9). Most call notes, however, are easy to locate. They start and stop abruptly and contain a wide range of pitches. If they are strung together they become even more easy to locate. The most complex call notes of birds are those uttered when an individual is taking flight or about to take off. If there is some value in maintaining contact with fellow flockmates a bird about to move gains if the others know where it is,

that it is moving or intends to do so, and where it is going. Some of these flight calls are very long and complex. In fact, the flight call complex is the longest vocalization given by male redwings, greatly exceeding the length of the song (fig. 11.6).

What Are Blackbirds Communicating?

When it was believed that the only function of signals was to communicate information accurately to other individuals, the vocalizations of a species were analyzed to determine the kinds of information communicated by each sound. I was personally very proud of a table I constructed on the vocalizations of red-winged, tricolored, and yellow-headed blackbirds. I assumed that if *I* could tell the species to which a bird belonged by hearing a sound, the sound carried information about species identity. If there was substantial individual variability in the structure of a sound, I assumed that it communicated individual identity. If the sound was uttered only by individuals of

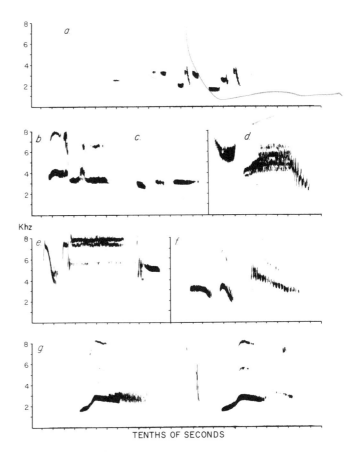

11.4 Blackbirds that defend large territories sing musical songs. Illustrated here are the songs of (a) western meadowlark; (b,c) scarlet-headed blackbird; (d) lesser red-breasted meadowlark; (e) red-breasted blackbird; (f) chopi; and (g) melodious blackbird.

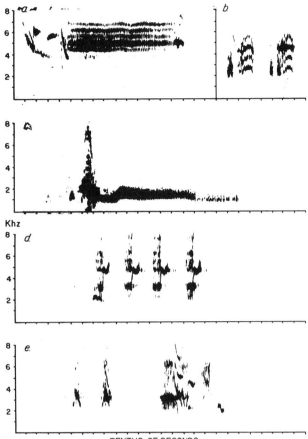

11.5 *Blackbirds that defend small territories sing low-pitched, nasal songs that are unattractive to the human ear. Illustrated here are (a) yellow-winged blackbird; (b) brown-and-yellow marshbird; (c) tricolored blackbird; (d) yellow-rumped cacique; and (e) russet-backed oropendola.*

one sex, it was assumed to carry sexual identity information. If the sound was normally elicited by some change in the environment, such as arrival of a predator, it was assumed to carry environmental information. If the sound provided information about the probable future behavior of the caller, it was credited with having social information.

What fascinates me now is that I never thought to ask, "What are they *not* communicating?" Much that could have been communicated was not. Nor did I really ask, "What does the individual gain by communicating that information and only that information?" Today I find these questions more interesting than the ones I asked then.

Vocalizations do communicate information about the identity of the signaler, and by appropriate experiments we can find out which parts of the signal contain the identifying information. This is easy to do because recorded vocalizations can be altered in whatever way we like and the modified sounds can be broadcast

without a real caller. This has been done with the songs of male redwings. These songs consist of three basic parts—an introductory syllable, a middle syllable, and a terminal trill (see fig 11.2). Songs can be cut up and spliced together in various orders to produce new songs. If the order of the syllables conveys species identity, this can be inferred from responses when the whole song is played but the order of the syllables is altered. The necessity of a particular syllable can be demonstrated by observing responses to songs with that part deleted.

Male redwings are ideal experimental subjects because they respond to taped songs played in their territories by approaching the loudspeaker and giving song spreads in front of it. The intensity of this response provides a good measure of whether or not they recognize the song as a threat. Male redwings do not care about the ordering of the syllables. A song consisting entirely of terminal trills, provided it is at least half as long as a normal song, gets the same response as a normal song. A song composed of a series of introductory or middle syllables, or a combination of them, even if it is the same length of a normal song, does not elicit much response. These parts of the song presumably have other functions, but they do not carry the information that says, "I am a male red-winged blackbird defending a territory."

Birds are highly mobile and frequently move into and out of sight of one another. They often do not know exactly where their mates, offspring, or flockmates are, but it may be mutually beneficial to maintain contact. Sounds can do so much more easily than visual displays (fig. 11.7).

Singing Together

Duetting is one of the most interesting forms of vocal communication among birds. In a number of species, pairs sing complicated duets in which each member

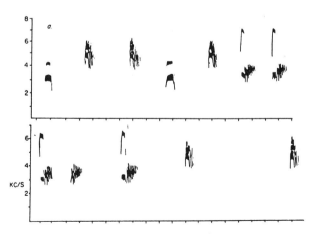

11.6 *The flight call complex of the male redwing.*

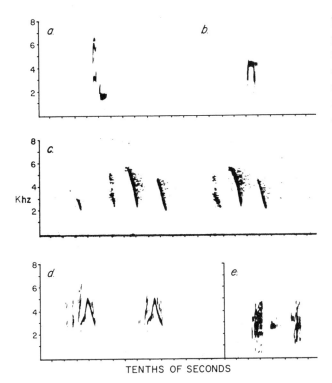

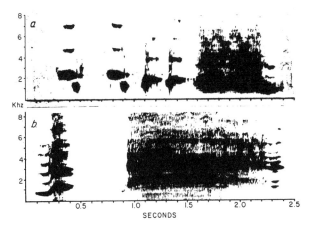

11.7 Some calls used by blackbirds to help maintain contact among individuals in flocks: (a) "seechuck," red-winged blackbird; (b) "cheet," redwing; (c) "chips," Bolivian blackbird; (d) "chips," brown-and-yellow marshbird; and (e) "chips," tricolored blackbird.

sings a different part of the duet and the timing of their utterances is precisely regulated. The most famous duetter among the blackbirds is the melodious blackbird of Central America, whose beautiful duetting is one of its most conspicuous activities, in part giving rise to its name. Because many duetting birds live in dense, brushy, tropical habitats, originally it was thought that the duets of most birds functioned to keep the mates in contact. Most duetters, however, sing when members of the pair are sitting side by side. Melodious blackbirds are usually within a foot of one another when they duet. Duets are directed to other pairs and to wandering individuals.

Duetting is rare among birds of high latitudes but is common among tropical species that remain permanently paired on territories. Widowed individuals remain on their territories and advertise for a replacement mate. If each of the individuals sings a different part of the duet, the singing of a solitary bird can indicate that the pair no longer exists and which member remains.

Even though duetting is rare among temperate zone birds, members of a pair may respond to one another's calling. A common grackle is much more likely to call right after its mate has called than at any other time, and they often chatter together. During the nest-

building and egg-laying periods the movements of the pair are highly coordinated, the male following the female. The male is more likely to follow if the female calls when she leaves the nest area than if she leaves silently. More precisely timed duets may have evolved from these less exact exchanges between members of a pair.

Why So Many Different Songs?

Field guides usually describe a single song type for a species, but males of most birds sing a number of different types of songs. For a number of years investigators tried to find differences in how the song types were used, that is, to discover unique messages that went with each one. For the most part this effort failed. Different song types are sometimes used in different circumstances. Male yellowheads have two song types, the first a somewhat musical song directed primarily at birds at long distances, and the other, a nasal song accompanied by an asymmetrical song spread, usually directed toward birds very close to the singer (fig. 11.8). Most birds, however, use all of their song types in all circumstances. There does not seem to be a special message associated with each one. Male redwings at Yellowwood Lake, Indiana, sing from two to eight songs each; any of the songs can be used in any situation.

Several hypotheses have been suggested to explain the use of many types of song by one individual. One argues that dominant birds can be recognized by their distinctive songs and that lower-ranking birds continually attempt to mimic them. The dominant birds would then benefit by having many different songs and changing them regularly so that it would be difficult for subordinates to mimic them. Although birds do regularly mimic the songs of neighbors and develop

11.8 Two song types of the male yellow-headed blackbird: (a) uttered primarily when the recipient is far away; (b) usually used when the recipient is close.

local dialects that sometimes change, no good tests of this hypothesis have yet been attempted.

The Beau Geste hypothesis, which has been tested, states that a bird singing many different song types gives the impression to listeners that a number of different individuals are present. This should discourage further exploration by newcomers, and a bird with a larger repertory should have fewer territorial challenges to deal with than one singing fewer song types. A Beau Geste bird should tend to change songs when it changes perches, adding to the impression that more individuals are present. It should also be possible to measure the effectiveness of large repertories in defending a territory.

The male redwings of Yellowwood Lake sing their songs in bouts of about twenty repetitions of a single song type before shifting to another. Ken Yasukawa found that they were more likely to change songs when they changed perches than they would have if changes were random. He also made tapes of one to eight song types and played them to territorial male redwings. At first the males responded strongly to the tapes, approaching the speakers and displaying, but they soon became habituated to the sounds. As predicted by the Beau Geste hypothesis their responses dropped off fastest to tapes consisting of only one song type, and barely at all to tapes consisting of four or more song types. Additional support for the hypothesis came from experiments in which territorial males were removed and the territories were "defended" by loudspeakers broadcasting repertories of different sizes. Territories defended by speakers playing complete repertories experienced less trespassing by neighbors than territories defended by speakers playing single song types.

According to still another hypothesis, there is a cost to producing many different song types and only old and experienced males can afford to do so. Thus, singing many different song types could serve to signal resource holding power. The results of Yasukawa's experiments are actually consistent with that hypothesis too, but there is, as yet, no evidence that it is really more costly to produce several different song types than just one. Talk is cheap but it does cost something and perhaps it costs more than we now realize.

Among bobolinks, individuals within a population frequently share song types unique to their locality. Most males sing two distinctive song types, and these dominant songs are not sung by males in other colonies as close as twenty to twenty-five kilometers away. Little is known about the stability of these song dialects, but males in two colonies in Minnesota sang the same song types in 1974 as they had in 1973. Young males probably learn the song type of their colony during their first breeding season, because bobolinks have stopped singing by the time the young of the year have hatched and do not sing again during the autumn.

Some of the best information about local dialects and the rate at which they change has been gathered by Francisca Feekes on the colonial yellow-rumped cacique in Surinam. Cacique colonies are formed by groups containing individuals of both sexes. In small colonies, founded by a single group, breeding is highly synchronous, but in larger colonies, in which additional groups move in, breeding is asynchronous. Each colony of caciques, regardless of its size, has its own distinctive song type or types. Every male in the colony sings those same songs, which are distinct from those of other colonies, even from those within earshot. In most colonies there are one or two song types, but sometimes there are three or even four.

In Surinam the colonies are deserted during the rainy season and occupied again during the dry season when breeding occurs. The same trees are often occupied year after year, and frequently by the same birds. Nonetheless, in a new breeding season, the songs differ from those used the previous year in the same tree. Even within a breeding season song types sometimes change. This change is always associated with the arrival of a new group of breeders, but songs do not always change when new groups arrive. Unfortunately, we do not know where the new birds come from and why they are sometimes imitated and sometimes not.

Feekes witnessed the changeover of songs in one colony. Song types had been constant for several weeks when a new group of females arrived and started nest building. At the same time a new song was heard, presumably produced by newly arriving males, but since none of the males were color banded this is not certain. Then there followed a period in which both old and new song types were sung, and Feekes observed many individual males singing all the song types currently being used. Finally, all males ended by singing two new song types that were slightly different both from one another and from the original songs.

These observations indicate that dialects are not adaptations to local environments but rather are the result of interactions among individuals competing with one another for space and mates. They also show that dialects differ in their stability in different species of blackbirds. All males may sing the same songs or they may sing different ones. Relationships among singing males are different in different species and we should not seek a single explanation of song types, imitation, or dialects. Given the rich array of social systems among blackbirds this is not surprising.

Table 11.1. **Number of Vocalizations and Displays of Different Icterids**

Species	Males		Females	
	Vocalizations	Displays	Vocalizations	Displays
Bullock's oriole	5	3	5	2
Jamaican blackbird	7	3	6	3
Oriole blackbird	6		6	
Yellow-headed blackbird	8	13	5	7
Yellow-winged blackbird	8	7	6	2
Red-winged blackbird	18	12	6	9
Tricolored blackbird	17	9	6	9
Yellow-hooded blackbird	6	5	4	2
Eastern meadowlark	9	7	8	8
Western meadowlark	7	7	6	8
Brown-and-yellow marshbird	8	5	8	5
Scarlet-headed blackbird	14	7	14	7
Bolivian blackbird	9	4	9	4
Melodious blackbird	5	7	7	8
Great-tailed grackle	5+	7	3	7
Boat-tailed grackle	5+	7	4	7
Common grackle	6	4	4	4
Brewer's blackbird	9	7	9	7

How Many Different Kinds of Vocalizations Do Blackbirds Have?

Even casual observers of birds know that some species do a lot of calling and singing and have many kinds of vocalizations while others are much quieter and give only a few kinds of calls. The number of distinct vocalizations that have been described for reasonably well-studied blackbirds are shown in table 11.1. These numbers are not precise because different workers combine or separate call types in slightly different ways, but the general patterns they reveal are probably accurate.

The number of vocalizations given by blackbirds ranges from only three or four up to nearly twenty. Among strongly polygynous species, males tend to have many more vocalizations than females, while in monogamous species the two sexes have similar numbers of calls. The colonial, promiscuous grackles are puzzling because males do not have many more vocalizations than females. This may be because male grackles do not hold spatial resources and therefore are not competing vocally with one another for those resources.

Concerning permanently paired tropical species, information is rather skimpy, but males and females appear to have about the same number and kind of vocalizations. Differences are found particularly among calls directly associated with precopulation and copulation. How many vocalizations are shared is an interesting indicator of the selective pressures on males

and females. Among redwings, for example, males have eighteen distinct vocalizations, females six. Of those six, four are shared with males, the other two being restricted to females. The shared ones are the basic alarm calls. Female redwings talk to other females more than they talk to males, and they are involved with complicated interactions concerning where they will place their nests, the timing of those nests, and where they will forage. Males talk a great deal with other males and with females at the time of their arrival in the spring. Later on, once females are incubating and feeding young, males have very little to do with them as individuals, although they may signal the presence of predators to all females in general.

Similarity in vocalizations reflects similarity of social roles. Among permanently paired tropical orioles, for example, males and females both defend the territory, seek replacement mates when necessary, and feed the young. We would expect them to have very similar vocalizations, as indeed they do. In species such as redwings and yellowheads, where males and females are engaged in different activities, it is not surprising that their vocalizations also show differences.

How Often Should One Utter Sounds?

A bird could have many different vocalizations and yet use all of them very infrequently. Or it could have just a few but use them all of the time. Noisiness need not be strongly correlated with repertory!

Energy considerations are probably not very important in determining the frequency of vocalizing because talk is cheap. Moreover, vocalizing is usually not very competitive with other uses of time because a bird can sing while actively foraging, merely interrupting its search briefly to utter a song before resuming its feeding. Territorial males of many small birds combine foraging and singing in this way. Steady bouts of song, uninterrupted by foraging, occur primarily at dawn and dusk—times when lighting conditions are too poor for efficient foraging. These are also times when potential new mates and intruding males are most likely to arrive.

Calling does attract attention to the caller—this is the presumed function of the sounds. However, calls can attract individuals the caller would prefer to avoid, such as predators. Moreover, for some vocalizations, particularly song, birds assume the most conspicuous positions they can, exposing themselves even more to attacks by predators. It is very difficult to measure predation rates on singing versus silent birds because predation is a relatively rare event and because the presence of an observer is likely to influence the activities of the predators.

There is indirect evidence, however, that predation risk may influence the rate of singing. In most monogamous birds, males sing vigorously when they first arrive and are establishing territories and until they attract their mates. Then song rate drops abruptly and strikingly. Among polygynous species, however, song rates do not change noticeably when a male gets his first female. Male redwings sing at about the same rate throughout the period when new females are arriving. One cannot tell from his singing rate whether a male already has females or how many. In many birds singing stops abruptly when a hawk is spotted. These observations suggest that there is a risk to singing and that birds quickly cease when the disadvantages outweigh the advantages.

How Do Blackbird Vocalizations Develop?

All birds are born virtually voiceless. Soon, however, they utter a weak begging note when an adult arrives at the nest with food. This note gradually increases in strength and, about the time the bird leaves the nest, changes into a more racuous, more readily located fledgling begging call. Also during this time a scream develops, given when the nestling is actually handled or attacked. Adults respond vigorously to this scream by mobbing the predator.

Other vocalizations develop later, although in no blackbird has the precise timing of their appearance been examined. Typically, singing begins during the first autumn in temperate zone species, but some vocalizations are not used until the first breeding season. Whether they could be uttered sooner is

unknown because the conditions appropriate for their use do not arise.

The temporal pattern of first use of different vocalizations does not tell us anything about how they develop. Does the full sound appear normally, without any opportunity for the individual to hear it? Or is a period of learning necessary? If so, how long a period of learning? Is there a special time when learning seems to occur? Why should learning occur at all? Why not know innately how to sing and call? There is no single answer to these questions because some birds sing perfect songs when reared in sound isolation from other members of their species while others fail to do so. Most of the simple call notes of birds are strongly genetically fixed, and develop in their proper form whether or not the bird has had a chance to hear others give those calls.

The advantage of not having to learn one's song is that the appropriate, species-specific song will be sung whatever the experiences of the developing individual. However, a genetically fixed song may be difficult to modulate and vary as conditions change. Learning one's song has the disadvantage that one may fail to hear the right song and, hence, never learn how to sing, or learn later than would be desirable. Also an individual may learn the wrong song, a very serious error if this prevented it from breeding. Even if the learner does hear the correct song during its learning period, it will inevitably be exposed to many other types of songs that it should reject in favor of the correct one.

Extensive experiments have shown that European chaffinchs, which do have to learn their songs, are resistant to learning incorrect songs but quickly learn a chaffinch song after a very brief exposure to it. Some such predisposition to learn is an essential feature of any species having a significant learned component to its song. Equal receptivity to all songs in its environment would result in absolute chaos.

For song learning to work at all, opportunities must exist to hear and learn the correct song at some time prior to the first breeding season when the individual will need to use it. In many birds, singing stops abruptly once pairs are formed, and there is no singing in the late summer, fall, or winter. In these species young males would have great difficulty in learning their songs in time, and genetic programing is much safer. In other species, however, songs continue longer in the summer, and there is also a period of renewed singing in the autumn as resident birds engage in some territorial behavior. In those species there are ample opportunities for young birds to learn their songs.

Rather little is known about song learning among blackbirds. Wesley Lanyon found that hand-reared meadowlarks given no opportunities to hear proper meadowlark songs nonetheless developed perfectly normal songs. They also learned and mimicked other songs, such as those of red-winged blackbirds, wood pewees, and yellowthroats. Whether or not they would have used these other songs in the field is questionable.

During the autumn of 1959, when I was studying some fall-breeding tricolored blackbirds in California, I removed nestlings from some nests when they were only a few days old and hand-reared them in my laboratory. They developed normal call notes, but their songs, though having the typical growly form of the song of the highly colonial tricolor, were highly abnormal in their overall form and structure (fig. 11.9). I should never have recognized the songs as belonging to a tricolor if I had only heard the recording.

Much of the preceding discussion about song learning is based upon the assumption that it is important to be able to match your song rather closely to those of neighbors, particularly socially dominant neighbors. But there is little evidence that deviating significantly from the local dialect is really important in determining the success of a territorial male. Recent experiments by Douglas Smith show how this problem can be approached and what the difficulties are in interpreting apparently simple experimental results. Smith trapped territorial male redwings, anaesthetized them, and then sectioned their hypoglossal nerves (the nerve activating the muscles of the tongue) on both sides. The control birds received identical treatments, including having their hypoglossal nerves exposed, but without being cut. Birds recovered from the operation rapidly and were released back on their territories within an hour of their first capture. The males with severed nerves sang very abnormal songs and showed signs of having respiratory problems which may also have affected their subsequent behavior. Despite these

problems only one of the fifteen experimental males lost his territory during that season while three of the fifteen control males did. Five of the experimentals but none of the controls returned the next year. Smith made recordings of the songs of these returning males and played them to neighbors and to other redwings elsewhere who were highly unlikely to have heard their songs.

In 1973, the year following the operation, the experimental males had, on the average, 1.4 females per territory while normal males had 1.3 females. Experimental males sang at the same rate as normal males (4.86 songs/minute vs. 4.44 songs/minute). Nonetheless, none of the males tested during playback experiments responded to the songs of the experimental males, while they did respond to the songs of normal males. Normal redwings did not recognize the songs of the operated males as belonging to their own species, but the experimental males successfully held territories and attracted the same number of females as did normal males!

In later experiments Smith devocalized several males by cutting the membranes of their interclavicular air sacs. Males regained their voices after about two to three weeks of being mute, but during their silence they all suffered high rates of intrusion from neighbors and their territories shrank in size. Two of the three regained their full territories when their voices returned.

These experiments show clearly that for a male to be a successsful territory holder, a song similar to the normal one is not necessary, and suggest that females do not discriminate against males with odd songs. The redwing, however, is perhaps a poor species on which to perform such experiments because females apparently choose males primarily on the basis of their real estate. Also, we do not know if those females that *were* attracted copulated more than usual with males on nearby territories. Because of the great importance of prior residency in territoriality, we also need results on birds receiving the operations when they are yearlings and have not yet had any territory. If such devocalized yearlings are as successful in holding territories and attracting females as normal males, we would have to rethink the function of song in this species more than we are forced to do by the results we have so far.

The true functions of song remain an enigma. In some ways we are like attendees at an opera being sung in a foreign language. From the action on stage we can tell that mates are being selected and resources competed for, but the utterances of the participants cannot be followed in any detail. We know that the opera would be more meaningful if we could understand the words, but it is a rich experience nonetheless.

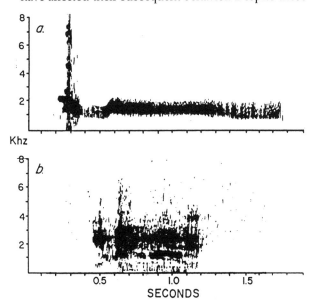

11.9 Song of the male tricolored blackbird: (a) normal song; (b) abnormal song of a bird reared in isolation.

12.1 *Two male Bullock's orioles fight vigorously for positions in trees that serve as courting and advertising centers.*

12. Of People and Blackbirds

As I write this final chapter in my home in Seattle, I look out on a bright green lawn, tall conifers and poplars, and many kinds of rhododendrons—not good icterid country. Occasionally in spring a Bullock's oriole takes a look around, evidently attracted by our tall Lombardy poplars, sings for a few days, and then decides that there are entirely too many conifers and not enough deciduous trees here. Brown-headed cowbirds, rather recent arrivals in Seattle, use our yard to parasitize some of the local song sparrows and rufous-sided towhees. Rarely, when I am working in the yard, I hear the distinctive "check" of a male redwing as he passes over, flying between the marshes along Lake Washington well below our home and some other, to me unknown, location. There is nothing in our yard to divert redwings from their course. Yet, as I write, my mind is full of images of blackbirds, of being roused from my sleeping bag at dawn in the Potholes by a western meadowlark proclaiming his territory just a few feet above my head, of flocks of millions of tricolored blackbirds streaming out of their wintering roosts in the Sacramento Valley of California to seek food in surrounding croplands, of eight brown-and-yellow marshbirds scolding us at a single nest where the three offspring probably had more attending adults than they knew what to do with.

I also imagine things that I have never experienced. I can imagine a blackbird making decisions about where to nest and think of a hypothesis to predict patterns I have not yet attempted to measure. I even imagine things that do not really exist. What would a blackbird species with three sexes be like? Would any of those sexes be like males and females as we know them or would all three be different? Why are there only two sexes and no more?

Science, like so many other forms of human behavior, requires the ability to hold ideas and images in our heads in the absence of the real thing, an ability to play with these ideas intellectually, and an ability to conceive of tests of those ideas. Scientific methods are nothing more than a set of rules about how to deal with hypotheses. The most important impact of the rules is that they help us devise tests that could falsify a hypothesis that is wrong. We all develop identities with our ideas and would prefer them to be correct rather than incorrect. To counter this bias, which might cause us to seek only favorable evidence, we need the discipline of rules.

Textbook descriptions of scientific methods leave out the most important part of science. The list of procedures starts with something like: "Given a hypothesis, then . . ." But where does the hypothesis come from? The real act of creativity is left out, but for a very good reason. It is ignored because we really do not understand where ideas and hypotheses come from. Something called imagination is involved, and some people have more of it than others. There are imaginative scientists and plodding scientists, just as there are imaginative and plodding painters.

Because we do not know quite what to do with imagination, we tend to concentrate on the rules of scientific methods when we describe science to other people. We are often embarrassed by the degree to which scientific results and theories are tentative statements. They represent progress reports that may be altered significantly in the near future, sometimes in ways that can only be dimly imagined today. The key to the process is that the rules do contain powerful self-correcting guidelines. Built into them are procedures to change our understanding and to alter current beliefs.

It is quite evident from the preceding chapters that for nearly every important question I have raised, we can offer only tentative progress reports. The simple questions first asked by children about birds cannot be given satisfactory answers. Why does the red-winged blackbird have red epaulets while the yellow-headed blackbird has a yellow head? Or, why are these colors and those of meadowlarks reversed in South American blackbirds? Not only can we not give definitive answers to these questions, we recently have changed our ideas about which questions are most pertinent. Some of these changes in perception did not occur because of experimental results but because of a general change in how we look at social interactions among animals. Yet, for all these uncertainties, science offers a progressive and self-correcting way of learning about the world and providing the basis for us to make informed judgments about our behavior and its consequences. For better or for worse, we have not discovered a more effective way of understanding the world around us than through science.

Why Study Blackbirds?

Given all of the important social problems confronting us, why would rational people spend as much time as I have studying blackbirds?

In part, the appeal of bird-watching is escape. Birds are beautiful animals, most of whom are active at times of day when we prefer to be active, and, although they do have severe problems of their own, we can somehow separate ourselves from them. When I observe a territorial challenge with vigorous fighting between two male icterids, I can calmly watch the progress of the fight to see who will win (fig. 12.1). I could never observe a fight between two people with the same dispassionate objectivity. Moreover, the problems of birds often do not really seem like problems. When I hear a male bird singing late in the spring, it probably means that he has lost his mate and is seeking a replacement, but I am likely to enjoy the song and be pleased that there is still some singing so late in the year.

The beauty of birds is one of their main attractions, but this raises the question, "Why do we find birds beautiful?" And if we answer, "Because they are colorful and sing pretty songs," this merely raises a new question. Why are colors and songs beautiful to us? Thinking seriously about bird-watching forces us to reflect on the basis of our own aesthetic sense. Unfortunately, there is no more difficult arena, in part because we tend to resist analysis of our aesthetic senses. Human emotions are among the most complex of biological phenomena. We believe that they are the result of the operation of our brains, coordinated in some way with other parts of our bodies, but at present our understanding of these properties is very limited.

Nonetheless, it is worth the effort to examine our aesthetic responses because some interesting patterns may be revealed (fig. 12.2). For example, people find the pure-toned songs of birds with large territories much more beautiful than the songs of colonial, densely packed species. A typical response to the songs of yellow-headed blackbirds is that of P. A. Taverner, who wrote in 1934:

The song of the Yellow-head—if song it can be called, as it lacks every musical quality—is like that of no other Canadian bird. Climbing stiff-leggedly up a reed or tule stalk the male, with wings partly raised, lowers his head as if to be violently ill, and disgorges a series of rough, angular consonants, jerkily and irregularly, with many contortions and writhings, as if their sharp corners caught in the throat and they were born with pain and travail. They finally culminate and bring satisfied relief in a long-drawn, descending buzz, like the slipping of an escapement in a clock spring and the consequent rapid unwinding and futile running down of the machinery. The general effect of the performance may be somewhat suggested by the syllables "klick-kluch-klee—klo-klu-klel—kriz-kri-zzzzzz-zeeeeee."

At this point I must make a personal confession. Although there is much to be said for Taverner's description, I have come to love the song. The sound of the first yellowhead in the spring evokes all sorts of positive emotional feelings—the result of years of working with the bird. The fact that I actually like the sound is, of course, one of the reasons why aesthetics is such a difficult subject. Nonetheless, I concede that the western meadowlark produces sounds of much higher musical quality.

People like pure tones and high notes. The hero and heroine of operas are nearly always tenors and sopranos, not baritones and altos. The bad guy in the opera is always a bass. Do we like high pure tones because the basic structure of our ears is attuned to such sounds? Do bird songs stimulate ancient receptor mechanisms or have we learned to associate them with the time of year when resources and weather conditions are improving, both for them and for us? Or is our preference simply the result of cultural conditioning? We don't know, but perhaps the study of our aesthetic responses to birds may be a source of useful information.

Curiosity is another reason we are attracted to birds. Curiosity may have killed some cats, but a cat without it does not have a bright future either. As with so many traits, too much or too little curiosity may hurt us. Curiosity is the motor that induces us to explore new frontiers, while imagination provides the steerage to our course. Most of science is done for the satisfaction of curiosity. The prime role of a teacher is to foster and encourage curiosity rather than to dispense facts. Nearly everything I have done with blackbirds has been driven by curiosity. I might have become interested in Gothic cathedrals instead, but I never encountered any in my formative years. My father did take me bird-watching, and that made all the difference.

The motivation for an activity and the benefits one gets from doing it may be quite unrelated. Exploring the other side of the mountain may provide access to new food supplies. Solving a scientific problem leads to publication and professional recognition. One pay-off from the study of other living organisms is a broadened understanding of ourselves. We share with blackbirds many of the same problems and needs. We seek suitable living habitats and are very much preoccupied with the pursuit, capture, and consumption of food. Going to the grocery store may be an aberrant form of hunting, but it does involve making decisions about which patches to visit and which prey to select once we are in the patch. We spend a great deal of time and money building nests, selecting mates, rearing offspring, and communicating with one another. We are very much concerned about male and female roles. We have developed the most elaborate communication system in the history of life and much of our daily lives is involved with communication. Paper-shuffling and

12.2 *The author contemplating the roots of aesthetics.*

telephone conversations occupy many hours every week.

The study of blackbirds can enlighten our self-awareness, but we must be wary of simple and loose analogies. Just because some blackbird is territorial does not mean that people are territorial any more than it means that all blackbirds must be territorial. Because some blackbird has a particular sex-role pattern does not say anything about what human sex-role patterns are or ought to be. The value of a comparative approach to human behavior lies not in the results but rather in the thought processes we have used and the types of hypotheses that have been formulated. Throughout the book I have shown how males and females respond differently to selective pressures because of basic differences in what it means to be a male or a female. I have shown how differences in the abundance and distribution of food can influence spacing and clumping patterns and where foraging occurs. I have indicated how social organization may influence the kind and number of displays and vocalizations given by members of a species. The same ideas can also be used to approach problems of human behavior.

During the past decade, a number of people have recognized the value of a comparative evolutionary approach and have begun to explore human behavior from this perspective. Some of the most interesting studies have involved the use of foraging theory to study hunting patterns in hunter-gatherer societies. Individuals in those groups make important decisions about where to hunt, the length of trip to prepare for, the type of hunting gear to take, and whether or not to go hunting at all. There are, of course, important differences between human hunters and other hunting animals. We communicate with one another about possible outcomes of different hunts. The extensive use we make of hunting tools means that we have to decide in

advance the type of prey we will go hunting for and which weapons we will take. Other species always have their gear with them. Nonetheless, the same cost-benefit arguments we use in studying blackbirds or other animals can also be used to approach our own hunting behavior. Attempts to use this way of thinking about human hunting groups is leading to important new insights into the structure of these societies. Unfortunately, however, hunting-gathering societies are rapidly disappearing and anthropologists of the future will, of necessity, study primarily cultures that have a more complex mixture of subsistence hunting and a market economy.

What Does the Future Hold for Blackbirds?

The blackbird family has evolved rather rapidly into the nearly one hundred species we find today. There is every reason to believe that evolutionary changes are still under way. The genus *Agelaius*, to which redwings belong, is one of those that are actively evolving. The tricolored blackbird almost certainly evolved from an isolated population of redwings west of the Sierra Nevada in California during the most recent ice age. The species is probably less than ten thousand years old. Redwings invaded Cuba twice in recent millennia. The first invasion led to the evolution of a species in which the females became darker than those on the mainland. Darkening of females has also occurred following the second invasion but has not proceeded so far (fig. 12.3). When redwings colonized Puerto Rico they also encountered conditions favoring the evolution of monomorphism in color. The time of this invasion is not known, but it was also probably quite recent.

Speciation has been active in recent years in some of the South American species as well. The most important events driving these changes have been the alternation of wet and dry periods that accompanied the advances and retreats of the glaciers at high latitudes. Associated with these changes in precipitation were dramatic expansions and contractions of the forests. During dry periods, evergreen, wet forests were restricted to rather small, isolated areas, mostly places where rainfall today is higher than in surrounding areas. Forest birds isolated in these areas developed adaptations to local conditions and diverged from one another. As forests expanded, the ranges of these birds also expanded until formerly isolated populations came into contact. In some cases the amount of differentiation that had occurred during isolation was not enough to cause barriers to interbreeding and a single species again occupied the whole area. In other cases, however, sufficient differences had accumulated that the populations did not interbreed.

12.3 Females, shown beneath their mates, have evolved to be black like the males in both the tawny-shouldered blackbird of Cuba (above) and the yellow-shouldered blackbird of Puerto Rico (below).

12.4 The red-bellied grackle of Colombia is endangered because the tropical montane forests in which it lives are rapidly being cut down.

The well-marked but similar species of oropendolas of the subgenus *Gymnostinops* occupy ranges centered on the former wet forest refuges (see fig. 9.10). No more than one species occurs in any locality even though their ranges abut in a number of areas. These species probably evolved from one species whose populations became isolated in forest refuges in the recent past.

Human activities, which are causing rapid disappearance of wet forests over large expanses of the New World tropics, are also causing isolation of populations of blackbirds and many other species. This should lead to local differentiation as it has in the past, but there is a major difference. Today the loss of forest is not caused by a climatic event likely to reverse itself. Growth of human populations in the tropics virtually guarantees that forest destruction will continue into the next century until most of the forests of Central and South America will have been converted to other uses. There is little likelihood of a major re-expansion.

The fate of tropical forest icterids is to become increasingly restricted to smaller and smaller patches of forests. The smaller the range occupied by a population, the more likely it is to go extinct. Local populations are vulnerable to destruction by purely local events and such local extinctions are probably commonplace (fig. 12.4). Under normal conditions local populations are replenished by immigrants from nearby populations. If there are no nearby populations to provide immigrants, local extinctions become permanent. When the Panama Canal was built at the turn of the century, a hilltop in the Canal Zone was isolated by the rising waters of Gatun Lake and became Barro Colorado Island. This island is the home of the Smithsonian Institution's Tropical Research Institute. The birds on the island have been intensively studied for many years, and many extinctions of bird species have been documented. Some species have apparently been lost because, in the absence of continued disturbance, brushy habitats and young forests have become rare on the island. But some of the species that have become extinct are birds of mature forests which are now widespread on Barro Colorado Island.

We can expect similar patterns of extinction to occur among birds in the isolated fragments of forests that are likely to remain in South America by the turn of the century. As is the case with Barro Colorado Island, there will not be sufficient immigrants to make up for the losses. At Barro Colorado Island the source of colonists is actually very close, but many tropical forest birds do not fly across broad open expanses of water. Nor do many cross broad expanses of open fields.

Therefore, tropical forest birds in general, and icterids in particular, are likely to face increasing rates of extinction not matched by equal rates of formation of new species. Since most of the species of icterids occur in the tropics, we may currently be at that point when there are more species of blackbirds than there ever have been before or ever will be again in the future.

12.5 *This eastern meadowlark, gathering insect larvae to feed its nestlings, is typical of thousands of members of its species and many other blackbirds during the warmer months of the year.*

There are now nearly one hundred species of icterids, but our grandchildren are unlikely to be able to see that number.

Open-country blackbirds, however, are strongly favored by human activities. Red-winged blackbirds are far more abundant in North America now than at any time in the past. Many tropical open-country species have expanded in both range and population size during the past several hundred years. Most temperate zone icterids are birds of open and semiopen habitats so there is little reason to expect that they will be in any trouble in the near future. In fact, our attempts to control blackbird populations in some areas of the United States clearly demonstrate the resilience of these birds. They benefit from the way we change landscapes and will be with us for some time to come.

Island species of birds are especially prone to extinction because of their small ranges and population sizes. Blackbirds are well represented on the islands of the West Indies but many of these are already in danger of extinction. Some of these reductions are owing to the cutting of forests, but others are caused by additional factors. The shrinking of the breeding range of the yellow-shouldered blackbird on Puerto Rico has been caused by the invasion of the shiny cowbird.

Most icterids are resident or migrate relatively short distances. A few of the high-latitude breeders in North America migrate to the tropics and even farther, and are vulnerable to changes on both their temperate breeding grounds and their tropical wintering grounds.

The North American orioles, all of which migrate to the tropics, winter in disturbed areas and second growth, a type of habitat that has become more common than before and one that is likely to remain common. The bobolink may be more vulnerable. Its wintering grounds, the Argentina pampas, are undergoing extensive modification. The grasslands are being replaced by cropland and short-cropped pastures—habitats much less suitable for bobolinks.

Why Be Concerned about Extinction?

But why should we care about the loss of species? After all, there are millions of species of animals and over eight thousand species of birds alone. Surely the loss of part of them is not a serious matter, especially when we consider that massive extinctions have occurred in the past and the world does not seem all the worse for wear because of that. In fact, I would really prefer not to share the globe with some of the dinosaurs and other beasts that roamed the continents in the past. The massive extinctions of the past, however, occurred at rates much slower than those at which species may go extinct during the next hundred years. These earlier extinctions took place over thousands or millions of years. They seem rapid only because we think about them in terms of the billions of years over which life has existed. In that framework a thousand years is but a moment.

Ecological adjustments and evolution of species

often take much longer than a thousand years, and the consequences of nearly instantaneous extinction of a large number of species may be quite different from the same number of species being lost over longer time spans. The rate at which species are being lost is unknown because, for many groups of plants and animals, most of the tropical species have not yet even been described and named. Nonetheless, it is possible to make some estimates based on our knowledge of how rich in species different ecological communities are, how rapidly species tend to go extinct in small habitat patches, and how rapidly habitats are being reduced to small patches. Using these types of evidence, some biologists have estimated that as many as one million species of plants and animals may go extinct within the next fifty years if current trends continue! This would be a truly unprecedented evolutionary event.

Another unprecedented event is likely to accompany these losses. The process of speciation requires a large enough area for geographical isolates of a population to form, differentiate, and then come together again. By examining islands we can determine the smallest size within which a single species appears to have split and formed two. The size of island needed is remarkably large. For birds, the minimum size appears to be that of Madagascar. Small mammals have speciated on islands as small as Luzon and Cuba. For larger species of both those groups, the minimal area needed is still larger. None of the national parks and game reserves we have established or are likely to establish in the future are large enough.

A species can maintain a population in a small game reserve or park, but, if the size of that population is small, genetic variability is lost, deleterious effects of inbreeding appear, and the population cannot preserve its normal genetic structure. This altered structure means that continued evolution within the population cannot proceed as it would have in a larger area. We may be witnessing the end of normal evolutionary processes in many of the larger vertebrates that have relatively low population densities. The species that are the most striking and conspicuous to us are the ones most likely to suffer severely from habitat destruction. The evolution of insects will doubtless continue unabated.

The value of the many species that share the planet with us is only recently becoming appreciated. In fact, their monetary values have increased with the increasing complexity of technology because we have found new uses for many more products of other living organisms. About half of the prescriptions written today in the United States contain at least one chemical produced by a living organism, and the value of these products exceeds many millions of dollars annually. We continually search the world for wild species that can provide new genes to confer resistance of crop plants to pests, pathogens, and climatic changes. A single new gene may be worth millions of dollars. As affluence and leisure time increase, the recreational value of other species also increases. The main source of foreign exchange of countries like Kenya is tourism, and tourists don't fly to Kenya to see corn fields. It is elephants, rhinoceroses, giraffes, lions, leopards, ostriches, and flamingos that induce them to part with their dollars, pounds, marks, francs, and yen. The aesthetic value of interactions with other species is immense, though difficult to quantify.

Few of us would prefer a world without singing birds. Birds have traditionally provided us primarily with aesthetic and recreational values, but there have been some medical payoffs. The parasitic honeyguides of Africa, unlike most vertebrates, are able to digest wax. The resistance of the tuberculosis bacterium to conventional antibiotics resides in its waxy covering. Honeyguides have two micro-organisms in their guts, a micrococcus and a yeast that break down the wax to useful forms. These micro-organisms are being studied as part of our efforts to control bacteria with waxy coats.

I am not aware of any great medical benefits that have been looked for or derived directly from blackbirds. Blackbirds are, however, very much involved with agriculture. Most of the species are largely or entirely insectivorous during the breeding season, and their foraging activities reduce levels of potentially destructive insects in agricultural areas. In the United States, meadowlarks are especially likely candidates for playing this role (fig. 12.5). On the other side of the ledger, blackbirds can be destructive to crops (fig. 12.6). It is impossible to make up a reasonable balance sheet, nor is it obvious that doing so would really enlighten us very much.

Blackbirds and the Quest

My life has been deeply involved with blackbirds. I have friends who have devoted similarly irrational amounts of time to the study of other groups of organisms, and who have come to hold feelings similar to mine for icterids. No longer can I hear the song of a redwing without special vibrations traversing my spine. As spring approaches I can hardly wait until I wade into a marsh and find my first nest of the year, its bluish eggs with dark blotches resting gently in their cup of fine grasses. No longer can I be truly objective about the quality of the song of the yellowhead. There was insufficient time for me to develop the same depth of feeling about Argentine marshes, but never have I warmed so quickly to a new environment as I did to that one. Amidst familiar habitats I found a new set of species, obviously allied to those I knew best but sufficiently different to stimulate me to want to find out why they were different.

I did not really find out why they were different, but I know more about them than I did before. Science is a continual process of removing the covers of little black boxes, only to find still smaller black boxes inside. Part of the excitement of exploring the unknown comes from the knowledge that no matter how hard one searches, the unknown still remains.

The study of blackbirds or any other organism is not a suitable activity for people who need to know the final truth. As it was for Don Quixote, it is the quest that counts. I have had a joyous quest, and I intend to continue it. There will be pauses, deflections down other pathways, but I will probably never totally abandon blackbirds. As a result of this quest I see my own life and the lives of my fellow humans in a different perspective. The inspiration people seek from observing other organisms changes as people age and civilizations rise and fall but it never vanishes. Perhaps the most important things we can pass on to our children are the desire to enjoy plants and animals and opportunities to be surrounded by many different species. Life will be a bit poorer if fewer than ninety-four species of blackbirds survive with us into the twenty-first century.

12.6 In early autumn, redwings often feed in large numbers on the heads of ripening corn

152

Appendix A. *Scientific Roster of the* Icteridae

Biologists classify organisms using a hierarchical system. Low-level units contain very closely related species that share recent common ancestors. Progressively higher-level units include species whose first common ancestors are even further into the distant past. Each species is given two names. The first name, which is capitalized, is its genus (plural, genera). The second name, not capitalized, is its species name. For example, the western meadowlark *(Sturnella neglecta)* is one member of the genus *Sturnella* which also includes the eastern meadowlark *(Sturnella magna)* as well as three South American species. How many species should be included in a genus is a somewhat arbitrary decision because nature does not recognize genera. They are concepts that help us organize our thinking about organisms. To be useful, genera should, on average, contain a number of species, otherwise the genus contains no new information, but if a species is sufficiently different from all others it may usefully be placed in a genus by itself. Such is the case of the yellow-headed blackbird *(Xanthocephalus xanthocephalus)*, the sole member of its genus. On the other hand, many orioles are very similar to one another in size and structure. Ornithologists indicate that by placing all two dozen species into the genus *Icterus*.

Above the genus is the family, a cluster of genera whose members share more features in common than members of genera placed in other families. Blackbirds are all believed to share a recent common ancestry, and are, accordingly, included in a single family, the Icteridae. The diversity of its members attests to the great potential of the ancestral model.

subgenus *Psarocolius*	Casqued oropendola *(Psarocolius oseryi)* Band-tailed oropendola *(Psarocolius latirostris)* Crested oropendola *(Psarocolius decumanus)* Green oropendola *(Psarocolius viridis)* Dusky-green oropendola *(Psarocolius atrovirens)* Russet-backed oropendola *(Psarocolius angustifrons)* Chestnut-headed oropendola *(Psarocolius wagleri)*
subgenus *Gymnostinops*	Montezuma oropendola *(Psarocolius montezuma)* Chestnut-mantled oropendola *(Psarocolius cassini)* Para oropendola *(Psarocolius bifasciatus)* Black oropendola *(Psarocolius guatimozinus)* Olive oropendola *(Psarocolius yuracares)*
subgenus *Cacicus*	Yellow-rumped cacique *(Cacicus cela)* Red-rumped cacique *(Cacicus haemorrhous)* Scarlet-rumped cacique *(Cacicus uropygialis)* Golden-winged cacique *(Cacicus chrysopterus)* Selva cacique *(Cacicus koepckae)* Mountain cacique *(Cacicus leucoramphus)*—includes *chrysonotus* Ecuadorian black cacique *(Cacicus sclateri)* Solitary black cacique *(Cacicus solitarius)* Yellow-winged cacique *(Cacicus melanicterus)*
subgenus *Amblycercus*	Yellow-billed cacique *(Cacicus holosericeus)*

Epaulet oriole (*Icterus cayanensis*)—includes *chrysocephalus*
Yellow-backed oriole (*Icterus chrysater*)
Yellow oriole (*Icterus nigrogularis*)
Jamaican oriole (*Icterus leucopteryx*)
Orange oriole (*Icterus auratus*)
Yellow-tailed oriole (*Icterus mesomelas*)
Orange-crowned oriole (*Icterus auricapillus*)
White-edged oriole (*Icterus graceannae*)
Spot-breasted oriole (*Icterus pectoralis*)
Black-throated oriole (*Icterus gularis*)
Streak-backed oriole (*Icterus pustulatus*)—includes *sclateri*
Hooded oriole (*Icterus cucullatus*)
Troupial (*Icterus icterus*)
Baltimore oriole (*Icterus galbula*)
Bullock's oriole (*Icterus bullockii*)—includes black-backed oriole (*I.b. abeille*)
Orchard oriole (*Icterus spurius*)—includes *fuertesi*
Black-cowled oriole (*Icterus dominicensis*)—includes *prosthemelas*
Black-vented oriole (Icterus wagleri)
St. Lucia oriole (*Icterus laudabilis*)
Martinique oriole (*Icterus bonana*)
Montserrat oriole (*Icterus oberi*)
Black-headed oriole (*Icterus graduacauda*)
Bar-winged oriole (*Icterus maculialatus*)
Scott's oriole (*Icterus parisorum*)

Jamaican blackbird (*Nesopsar nigerrimus*)

Saffron-cowled blackbird (*Xanthopsar flavus*)

Oriole blackbird (*Gymnomystax mexicanus*)

Yellow-headed blackbird (*Xanthocephalus xanthocephalus*)

Yellow-winged blackbird (*Agelaius thilius*)
Red-winged blackbird (*Agelaius phoeniceus*)
Tricolored blackbird (*Agelaius tricolor*)
Yellow-hooded blackbird (*Agelaius icterocephalus*)
Tawny-shouldered blackbird (*Agelaius humeralis*)
Yellow-shouldered blackbird (*Agelaius xanthomus*)
Unicolored blackbird (*Agelaius cyanopus*)
Pale-eyed blackbird (*Agelaius xanthophthalmus*)
Chestnut-capped blackbird (*Agelaius ruficapillus*)

Red-breasted blackbird (*Leistes militaris*)

Peruvian red-breasted meadowlark (*Sturnella bellicosa*)
Lesser red-breasted meadowlark (*Sturnella defilippi*)
Long-tailed meadowlark (*Sturnella loyca*)
Eastern meadowlark (*Sturnella magna*)
Western meadowlark (*Sturnella neglecta*)

Yellow-rumped marshbird (*Pseudoleistes guirahuro*)
Brown-and-yellow marshbird (*Pseudoleistes virescens*)

Scarlet-headed blackbird (*Amblyramphus holosericeus*)

Red-bellied grackle (*Hypopyrrhus pyrohypogaster*)

Austral blackbird (*Curaeus curaeus*)
Forbes's blackbird (*Curaeus forbesi*)

Chopi blackbird (*Gnorimopsar chopi*)

Bolivian blackbird (*Oreopsar bolivianus*)

Velvet-fronted blackbird (*Lampropsar tanagrinus*)

Mountain grackle (*Macroagelaius subalaris*)
Golden-tufted grackle (*Macroagelaius imthurni*)

Cuban blackbird (*Dives atroviolaceus*)
Melodious blackbird (*Dives dives*)
—includes *warszewiczi* of Peru

subgenus *Cassidix*

Boat-tailed grackle (*Quiscalus mexicanus*)
Great-tailed grackle (*Quiscalus major*)
Slender-billed grackle (*Quiscalus palustris*)—extinct
Nicaraguan grackle (*Quiscalus nicaraguensis*)

subgenus *Quiscalus*

Common grackle (*Quiscalus quiscula*)

subgenus *Holoquiscalus*

Greater Antillean grackle (*Quiscalus niger*)
Carib grackle (*Quiscalus lugubris*)

Rusty blackbird (*Euphagus carolinus*)
Brewer's blackbird (*Euphagus cyanocephalus*)

Bay-winged cowbird (*Molothrus badius*)
Screaming cowbird (*Molothrus rufoaxillaris*)
Shiny cowbird (*Molothrus bonariensis*)
Red-eyed cowbird (*Molothrus aeneus*)
Brown-headed cowbird (*Molothrus ater*)

Giant cowbird (*Scaphidura oryzivora*)

Bobolink (*Dolichonyx oryzivorus*)

Appendix B. *Habitat, Diet, Plumage, and Social Organization of Blackbirds*

Species	Differences in Wing Length between Males and Females	Habitat Breeding	Habitat Nonbreeding	Seasonal Movements	Diet Breeding Season	Diet Nonbreeding Season	Spacing Breeding	Spacing Nonbreeding	Where Pair Bond Is Formed	Mating Relationship	Helpers at Nest	Role of Sexes Male	Role of Sexes Female	Nest Site	Immature Plumage	Bill	Crown	Head	Back	Breast	Belly	Legs	Flash Colors
Psarocolius oseryi	30%	F	F	N	O	O	L	F	O			G	N,I,F	Tc	0	I	C	C	C	Y	C	B	T,U
P. latirostris	35%	F	F	N	O	O	L	F	O			G	N,I,F	Tc	0	I	C	C	B	B	B	B	T,U
P. decumanus	28%	E	E	N	O	O	L	F	O	P	–	G	N,I,F	Tc	0	I	B	B	B	B	B	B	T,U
P. viridis	27%	F	F	N	O	O	L	F	O			G	N,I,F	Tc	0	I	G	G	G/C	G	C	B	T,U
P. atrovirens		F	F	N	O	O	L	F	O			G	N,I,F	Tc	0	G	G	G	G/C	G	G	B	T,U
P. angustifrons	15%	F	F	N	O	O	L	F	O	P	–	G	N,I,F	Tc	0	Y	C	C	C	C	C	B	T,U
P. wagleri	29%	E	E	N	O	O	L	F	O	P	–	G	N,I,F	Tc	0	I	C	C	B	C	B	B	T,U
P. montezuma	25%	E	E	N	O	O	L	F	O	P	–	G	N,I,F	Tc	0	B O	B	B	C	B	B	B	T,U
P. cassini		F	F	N	O	O	L	F	O			G	N,I,F	Tc	0	B R	B	B	C	B	B	B	T,U
P. bifasciatus		F	F	N	O	O	L	F	O			G	N,I,F	Tc	0	B R	B	B	C	C	B	B	T,U
P. guatimozinus	23%	E	E	N	O	O	L	F	O			G	N,I,F	Tc	0	B Y	B	B	B/C	B	B	B	T,U
P. yuracares	15%	F	F	N	O	O	L	F	O			G	N,I,F	Tc	0	Y	Y	Y	C	Y	C	B	T,U
Cacicus cela	23%	E	E	N	O	O	L	F	O	P	–	G	N,I,F	Tc	0	I	B	B	B,Y	B	B	B	W,R,U
C. haemorrhous	22%	E	E	N	O	O	L	F	O	P	–	G	N,I,F	Tc	0	I	B	B	B,R	B	B	B	R
C. uropygialis	12%	F	F	S	I	I	T	F		M		G,F	N,I,F	Th	0	I	B	B	B,R	B	B	B	R
C. chrysopterus	13%	F,R	F	S	I	I	S	S								I	B	B	B,Y	B	B	B	W,R
C. koepckae	5%	F	F	S	I	I	S	S								I	B	B	B,Y	B	B	B	R
C. leucoramphus	21%	F	F	S	I	I	S	S								I	B	B	Y	B	B	B	W,R
C. sclateri		F	F	S	I	I	T	S							0	I	B	B	B	B	B	B	O
C. solitarius	10%	E	E	S	I	I	T	S		M					Th	0	I	B	B	B	B	B	B	O
C. melanicterus	15%	S,B	S,B	N	I	I	S,L	F		P		G	N,I,F	Tc	0	I	B	B	B,Y	B	B	B	W,R,U T,U
C. holosericeus	10%	E	E	S	I	I	T	S		M			N,I,F	S	0	Y	B	B	B	B	B	B	O
Icterus cayanensis	3%	E	E	S	I	I				M				T		B	Y	B	B	B	B	B	W
I. chrysater	14%	E	E	S	I	I				M				T	1	B	Y	Y,B	Y	B,Y	Y	B	O
I. nigrogularis	9%	D	D	S	I	I				M				Tc		B	Y	Y	Y	Y,B	Y	B	W
I. leucopteryx	7%	E	E	S	I	I,N				M				T	1	B	Y	Y	Y	B	Y	B	R,W
I. auratus	1%	B	B	S	I	I,N				M				T	0	B	O	O,B	O	B,O	O	B	W
I. mesomelas	3%	E	E	S	I	I				M	F	N,I,F		S,T	1	B	Y	Y,B	B,Y	B,Y	Y	B	W,T
I. auricapillus	2%	E	E	S	I	I				M				T		B	O	B,Y	B,Y	B,Y	Y	B	W,R
I. graceannae	4%	D	D	S	I	I				M				T		B	Y	B,Y	B,Y	B,Y	Y	B	W,T
I. pectoralis	5%	E,S	E,S	S	I	I	T	F		M	F	N,I,F		Tc	1	B	O	O,B	B,O	B	O	B	W
I. gularis	8%	E	E	S	I	I	T	S		M	F	N,I,F		Tc	1	B	O	O,B	B,O	B,O	O	B	W
I. pustulatus	8%	E,S	E,S	S	I	I				M				Tc	1	B	O	O,B	B,O	B,O	O	B	W
I. cucullatus	5%	E,U	E,U	P	I	I	T	F	T	M				Tc	2	B	O	O,B	B,O	B,O	O	B	W
I. icterus	6%	D,S	D,S	S	I	I,N				M				Tc	0	B	B	B	B,O	B	O	B	W,R
I. galbula	7%	F,R,U	F,R,U	M	I	I,N	T	F	T	M	–	D,F,G	D,N,I,F,G	Tc	2	B	B	B	B,O	O	O	B	W,T
I. bullockii	7%	F,R,U	F,R,U	M	I	I,N	L	F	T	M		F,G	N,I,F,G	Tc	2	B	B	O,B	B	B,O	O	B	W,T
I. spurius	6%	E,U	E,U	M	I	I	T,L	F	T	M		D,F,G	D,N,I,F,G	S,H,Th	2	B	B	B	B	B,C	C	B	W
I. dominicensis	5%	F	F	S	I	I	T	S		M			N,I,F	T	1	B	B	B	B,Y	B	Y	B	W
I. wagleri	7%	E	E	S	I	I				M				T	1	B	B	B	B,Y	B,Y	Y	B	W
I. laudabilis	0%	E	E	S	I	I				M				T	1	B	B	B	B,O	B	O	B	R
I. bonana		E	E	S	I	I				M				T	0	B	C	C	C,B	C	O	B	R

Species	Differences in Wing Length between Males and Females	Habitat Breeding	Habitat Nonbreeding	Seasonal Movements	Diet Breeding Season	Diet Nonbreeding Season	Spacing Breeding	Spacing Nonbreeding	Where Pair Bond Is Formed	Mating Relationship	Helpers at Nest	Role of Sexes Male	Role of Sexes Female	Nest Site	Immature Plumage	Bill	Crown	Head	Back	Breast	Belly	Legs	Flash Colors
Icterus oberi	9%	.E	E	S	I	I				M				T	0	B	B	B	B	B	Y	B	W,T
I. graduacauda	5%	F,R	F,R	S	I	I				M				T	1	B	B	B	Y	B,Y	Y	B	W
I. maculialatus		E	E	S	I	I				M				T	0	B	B	B	B,Y	B,Y	Y	B	R,W
I. parisorum	6%	D	E	M	I,N	I,N	T	F		M		F	N,I,F	S,Tc	2	B	B	B	B,Y	B	Y	B	W,T,U
Nesopsar nigerrimus	5%	F	F	S	I	I	T	T	?	M	–	F	I,F	Tc	1	B	B	B	B	B	B	B	O
Xanthopsar flavus	5%	M	T,M	N	I	I,G								E	2	B	Y	Y	B	Y	Y	B	R,W,U
Gymnomystax mexicanus	0%	S,B,L	S,B,L	N	I	I,G		F		M		G,F	N,I,F	Th	0	B	Y	Y	B	Y	Y	B	W,O
Xanthocephalus xanthocephalus	19%	M	M,C	P	I	I,G	G	F	T	P	–	D,F	N,I,F	E	2	B	Y	Y	B	Y,B	B	B	W
Agelaius thilius	9%	M	M,C	P	I	I,G	L	F	A	M	–	G,F	N,I,F	E	2	B	B	B	B	B	B	B	W,U
A. phoenicus	18%	M,T	M,C	P	I,G	I,G	G	F	T	P	–	D,F	N,I,F	E	2	B	B	B	B	B	B	B	W
A. tricolor	12%	M,T	M,C	N	I,G	I,G	D	F	T	P	–	D,F	N,I,F	E	2	B	B	B	B	B	B	B	W
A. icterocephalus	12%	M	M,T	N	I	O	T,L	F	T	P	–	N,D,F	I,F	E	2	B	Y	Y	B	Y	B	B	O
A. humeralis	8%	S,U	S,U	S	I	O		F								B	B	B	B	B	B	B	W
A. xanthomus	10%	B,M,U	S,U	N	I	O	L	F	A	M	–			Th,S	1	B	B	B	B	B	B	B	W
A. cyanopus	8%	M	M	N	I	O	L	F						E	2	B	B	B	B	B	B	B	O
A. xanthophthalmus	14%	M	M											E		B	B	B	B	B	B	B	O
A. ruficapillus	13%	M	M	N	I	G	L	F		P		D,N?	I,F	E		B	C	B	B	B	C	B	O
Leistes militaris	9%	T,L	T,L	S	I	I,G	T	F	T	M,P		T	N,I,F	G	1	B	B	B	B	R	B	b	W
Sturnella loyca	7%	T,L	T,L	P	I	O	T	F		M				G	0	b	B	B	B	R	R	b	U,T
S. defilippi	8%	T,L	T,L	P	I	O	T	F		M				G	0	b	B	B	B	R	R	b	U,T
S. bellicosa	13%	L	L	S	I	O	T	F		M				G	0	b	B	B	B	R	R	b	U,T
S. magna	10%	T,W,C	T,C	P	I	I,G	T	F	T	M,P	–	D,F	N,I,F	G	0	b	b	b	b	Y	Y	b	U,T
S. neglecta	11%	L	L,C	P	I	I,G	T	F	T	M,P	–	D,F	N,I,F	G	0	b	b	b	b	Y	Y	b	U,T
Pseudoleistes guirahuro	8%	W	W,C	P				F							0	B	b	b	G	G	Y	B	W,R,U
P. virescens	0%	W	W,C	N	I	I,G	L	F	A	M	+	G,F	N,I,F	H	0	B	b	b	G	G	Y	B	W,U
Amblyramphus holosericeus	6%	M	M,C	N	I	I,G	T	F	A	M	–	G,F	N,I,F	E	1	B	R	R	B	R,B	B	B	O
Hypopyrrhus pyrohypogaster	14%	F	F													B	B	B	B	B	R	B	U
Curaeus curaeus	2%	E,B,D	E,B,D,C	N	I,N	I,G				M	+			Th	0	B	B	B	B	B	B	B	O
C. forbesi															0	B	B	B	B	B	B	B	O
Gnorimopsar chopi	6%	S,B,R	S,B,R	S	I	O								Th		B	B	B	B	B	B	B	O
Oreopsar bolivianus	5%	D	D	S	I,F	I,F	G	G	T	M	+			C	0	B	B	B	B	B	B	B	W
Lampropsar tanagrinus	10%	R	R	S	I	I										B	B	B	B	B	B	B	O
Macroagelaius subalaris		F	F	S											0	B	B	B	B	B	B	B	U
M. imthurni	3%	F	F	S											0	B	B	B	B	B	B	B	U

Species	Differences in Wing Length between Males and Females	Habitat Breeding	Habitat Nonbreeding	Seasonal Movements	Diet Breeding Season	Diet Nonbreeding Season	Spacing Breeding	Spacing Nonbreeding	Mating Where Pair Bond Is Formed	Mating Relationship	Helpers at Nest	Role of Sexes Male	Role of Sexes Female	Nest Site	Immature Plumage	Bill	Crown	Head	Back	Breast	Belly	Legs	Flash Colors
Dives dives	11%	E,R S,B	E,R S,B	N	I	O	T	F		M		D,G,F	D,N,I,F	Th	0	B	B	B	B	B	B	B	O
D. atroviolaceus	6%														0	B	B	B	B	B	B	B	O
Quiscalus mexicanus	20%	S,U	S,U,C	N	O	O	L	F		P		G	N,I,F	Tc	1	B	B	B	B	B	B	B	O
Q. major	22%	S,U	S,U,C	N	O	O	L	F		P		G	N,I,F	Tc	1	B	B	B	B	B	B	B	O
Q. palustris (extinct)	22%	M	M	S			L	F						E	1	B	B	B	B	B	B	B	O
Q. nicaraguensis	22%	M	M	S	O	O	L	F		P			N,I,F	E	1	B	B	B	B	B	B	B	O
Q. quiscula	11%	E,U	E,U,R,W	P	O	O	L	F		M	–	G,F	N,I,F	Th	1	B	B	B	B	B	B	B	O
Q. niger	14%	S,U	S,U,C	S	I	I	L	F		M	–		N,I,F	Th	1	B	B	B	B	B	B	B	O
Q. lugubris	13%	S,U	S,U,C	S	I	I	L	F		M	–	G,F	N,I,F	Th	1	B	B	B	B	B	B	B	O
Euphagus carolinus	5%	b	L,W,C	M	I	O	L	F	T	M		G,F	N,I,F	E	1	B	B	B	B	B	B	B	O
E. cyanocephalus	8%	R,B	L,W,C	P	I	O	G	F	A	M	/	G,F	N,I,F	E,Th,S,G	1	B	B	B	B	B	B	B	O
Molothrus badius	5%	S,B	S,B,C	N	I	O		F		M	+	F	I,F	Tc	0	b	b	b	b	b	b	b	W
M. rufoaxillaris	6%	S,B	S,B,C	N	I	O	T	F		M		O	O		1	B	B	B	B	B	B	B	U
M. bonariensis	12%	S,B F,E	S,B,C	P	I	O	T	F		M		D	O		1	B	B	B	B	B	B	B	O
M. aeneus	11%	B,L	B,L,C	N	I	O		F				O	O		1	B	B	B	B	B	B	B	O
M. ater	10%	F,E S,M U	L,C	P	I	O	T	F		M		D	O		1	B	b	b	B	B	B	B	O
Scaphidura oryzivora	24%	F,E	F,E	N	O	O		F				O	O		2	B	B	B	B	B	B	B	O
Dolichonyx oryzivorus	11%	T,W	T,W,C	M	I	G	T	F	T	P	–	D,F	N,I,F	G	1	I	B	B,W	W	B	B	b	W,R

CODE FOR APPENDIX B

Habitat: F = forest; E = edge; R = riparian; S = savanna; B = bushland; T = tall grassland; L = low grassland; W = wet meadow; M = marsh; b = bog; s = swamp; D = desert scrub; U = parks and gardens; C = cropland.

Seasonal Movements: S = sedentary; N = nonbreeding nomadism within general breeding range; P = partially migratory; M = completely migratory (winter and breeding ranges totally distinct).

Diet: F = frugivorous; N = nectarivorous; G = granivorous; I = insectivorous; O = omnivorous.

Spacing: T = territorial; I = isolated but not territorial; G = grouped territories; L = loosely colonial; D = densely colonial; S = solitary; F = flocking.

Mating: *Where pairs formed:* A = away from breeding area; T = on territory; C = in colony; O = no real pairs formed. *Mating relationship:* M = monogamous; P = polygynous or promiscuous. *Helpers at nest:* + = regularly; / = occasionally; − = very rarely or never.

Role of Sexes: D = defends territory; N = builds nest; I = incubates; F = feeds both nestlings and fledglings; G = guards nest; O = no role because is a brood parasite.

Nest Site: Tc = trees, conspicuous; Th = trees, hidden; S = shrubs; H = herbaceous vegetation above ground; G = on ground; E = emergent aquatic vegetation; C = holes in cliff; − = does not build nest.

Immature Plumage: 0 = none; 1 = first fall and winter only, both sexes; 2 = through first breeding season, males only.

Adult Plumage: B = black or dark gray; W = white; b = brown; Y = yellow; O = orange; I = ivory or light gray; C = chestnut; R = red; G = green or olive; S = streaked. *Flash colors:* O = none; W = epaulets or wing bars; R = rump; T = outer tail feathers; U = under wings or tail.

References

Information used in this book was drawn from many sources. To aid readers interested in consulting the original references, the following list is grouped by chapter. For each chapter, citations are provided for general books and papers dealing with the subject matter of the chapter and, in addition, for those studies whose results were specifically cited in the test.

1. Of Blackbirds and Black Birds

Baker, V. R. 1978. The Spokane flood controversy and the Martian outflow channels. Science 202:1249-56.

Bent, A. C. 1965. *Life Histories of North American Blackbirds, Orioles, Tanagers, and Allies.* Dover Publications, New York.

Bretz, J. H. 1959. Washington's Channelled Scablands. State of Washington, Division of Mines and Geology, Bulletin 45:1-57.

Clark, T. H. and C. W. Stern. 1968. *Geological Evolution of North America.* Ronald Press, New York.

Harrington, H. J. 1962. Paleographic development of South America. Bulletin of the American Association of Petroleum Geologists 46:1773-1814.

Sutton, O. G. 1961. *The Challenge of the Atmosphere.* Harper and Brothers, New York.

van Tyne, J., and A. J. Berger. 1959. *Fundamentals of Ornithology.* John Wiley & Sons, New York.

Wallace, J. M., and P. V. Hobbs. 1957. *Atmospheric Science.* Academic Press, New York.

2. Gains from Gaping

Angell, T. 1978. *Ravens, Crows, Magpies, and Jays.* University of Washington Press, Seattle.

Beecher, W. J. 1951. Adaptations for food getting in the American blackbirds. Auk 68:411-40.

Lorenz, K. Z. 1949. Uber die Beziehungen zwischen Kopfform und Zirkelbewegung bei Sturniden und Icteriden. In: E. Mayr (ed.), *Ornithologie als biologische Wissenschaft.*

Orians, G. H. 1980. *Some Adaptations of Marsh-nesting Blackbirds.* Princeton University Press, Princeton, NJ.

3. Finding and Capturing Food

Alcock, J. 1973. Cues used in searching for food by red-winged blackbirds, *Agelaius phoeniceus.* Behaviour 46:174-88.

Campbell, R. W. 1974. Rusty blackbirds prey on sparrows. Auk 86:291-93.

Charnov, E. L. 1976. Optimal foraging, the marginal value theorem. Theoretical Population Biology 9:129-36.

Dolbeer, R. A. 1980. Blackbirds and corn in Ohio. U.S. Department of the Interior, Fish and Wildlife Service Resource Publication 136.

Fink, L. S., and L. P. Brower. 1981. Birds can overcome the cardenolide defense of monarch butterflies in Mexico. Nature 291:67-70.

Follett, W. I. 1957. Bronzed grackles feeding on emerald shiners. Auk 74:263.

Hamilton, W. J., Jr. 1951. The food of nestling bronzed grackles, *Quiscalus quiscula versicolor,* in central New York. Auk 68:213-17.

Helms, C. W. 1962. Red-winged blackbird killing a sharp-tailed sparrow. Wilson Bulletin 74:89-90.

Mason, J. R., and R. F. Reidinger. 1982. Observational learning of food aversions in red-winged blackbirds *(Agelaius phoeniceus).* Auk 99:548-54.

Meanley, B. 1971. Blackbirds and the southern rice crop. U.S. Department of the Interior, Fish and Wildlife Service Resource Publication 100.

Neff, J. A., and B. Meanley. 1957. Blackbirds and the Arkansas rice crop. Arkansas Agricultural Experiment Station Bulletin 584.

Orians, G. H. 1980. *Some Adaptations of Marsh-nesting Blackbirds.* Princeton University Press, Princeton, NJ.

———, and N. E. Pearson. 1979. On the theory of central place foraging. In: *Analysis of Ecological Systems,* D. J. Horn, R. D. Mitchell and G. R. Stairs (eds.). Ohio State University Press, Columbus, pp. 155-177.

Post, W., and J. W. Wiley. 1977. The shiny cowbird in the West Indies. Condor 79:119-21.

Pyke, G. H., H. R. Pulliam, and E. L. Charnov. 1977. Optimal foraging: a selective review of theory and tests. Quarterly Review of Biology 52:137-54.

4. Food and Blackbird Distribution

Case, N. A., and O. H. Hewitt. 1963. Nesting and productivity of the red-winged blackbird in relation to habitat. The Living Bird 2:7-20.

Charnov, E. L., G. H. Orians, and K. Hyatt. 1976. Ecological implications of resource depression. American Naturalist 110:247-59.

Ewald, P. R., and S. Rohwer. 1982. Effects of supplemental feeding on timing of breeding, clutch size, and polygamy in red-winged blackbirds *Agelaius phoeniceus.* Journal of Animal Ecology 51:429-50.

Lanyon, W. E. 1956. Ecological aspects of the sympatric distribution of meadowlarks in the north-central states. Ecology 37:98-108.

———. 1957. The comparative biology of the meadowlarks *(Sturnella)* in Wisconsin. Publication of the Nuttall Ornithological Club, no. 1.

———. 1962. Species limits and distribution of meadowlarks of the desert grasslands. Auk 79:183-207.

Lowther, P. E. 1975. Geographic and ecological variation in the family Icteridae. Wilson Bulletin 87:481-95.

Miller, R. S. 1968. Conditions of competition between redwings and yellow-headed blackbirds. Journal of Animal Ecology 37:43-61.

Orians, G. H., and G. Collier. 1963. Competition and blackbird social systems. Evolution 17:449-59.

———, and H. S. Horn. 1969. Overlap in foods and foraging of four species of blackbirds in the Potholes of central Washington. Ecology 50:930-38.

Payne, R. B. 1969. Breeding seasons and reproductive physiology of tricolored blackbirds and red-winged blackbirds. University of California Publications in Zoology 90:1-115.

Rohwer, S. A. 1973. Distribution of meadowlarks in the central and southern Great Plains and the desert grasslands of eastern New Mexico and west Texas. Transactions of the Kansas Academy of Sciences 75:1-19.

———. 1976. Species distinctness and adaptive differences in southwestern meadowlarks. Occasional Papers of the Museum of Natural History, University of Kansas 44:1-14.

Snelling, J. C. 1968. Overlap in feeding habits of red-winged blackbirds and common grackles nesting in a cattail marsh. Auk 85:560-85.

Voigts, D. K. 1973. Food overlap of two Iowa marsh icterids. Condor 75:392-99.

5. Gains from Grouping

Alexander, R. D., and G. Borgia. 1978. Group selection, altruism, and the levels of organization of life. Annual Review of Ecology & Systematics 9:449-74.

Brown, J. L. 1964. The evolution of diversity in avian territorial systems. Wilson Bulletin 76:160-69.

Caraco, T., S. Martindale, and H. R. Pulliam. 1980. Flocking: advantages and disadvantages. Nature 285:400-1.

Dyer, M. I. 1967. An analysis of blackbird flock feeding behavior. Canadian Journal of Zoology 45:765-72.

Hamilton, W. J., III, and K. E. F. Watt. 1970. Refuging. Annual Review of Ecology and Systematics 1:263-86.

Horn, H. S. 1968. The adaptive significance of colonial nesting in the Brewer's Blackbird (Euphagus cyanocephalus). Ecology 49:682-94.

Kenward, R. E. 1978. Hawks and doves: factors affecting success and selection in goshawk attacks on wood-pigeons. Journal of Animal Ecology 47:449-60.

Lanyon, W. E. 1956. Territory in meadowlarks, genus Sturnella. Ibis 98:485-89.

Orians, G. H. 1961. Ecology of blackbird (Agelaius) social systems. Ecological Monographs 31:285-312.

Schoener, T. W. 1968. Sizes of feeding territories among birds. Ecology 49:123-41.

Selander, R. K. 1958. Age determination and molt in boat-tailed grackles. Condor 60:355-76.

Smith, J. N. M. 1977. Feeding rates, search paths, and surveillance for predators in great-tailed grackle flocks. Canadian Journal of Zoology 55:891-98.

Wittenberger, J. F. 1978. The breeding biology of an isolated bobolink population in Oregon. Condor 80:355-71.

6. Nests and Nest Sites

Chapman, F. M. 1928. The nesting habits of Wagler's Oropendola (Zarhynchus wagleri) on Barro Colorado Island. Bulletin of the American Museum of Natural History 43:123-66.

Collias, N. E. 1964. The evolution of nests and nest-building in birds. American Zoologist 4:175-90.

———, and E. C. Collias. 1984. Nest Building and Bird Behavior. Princeton University Press, Princeton, NJ.

Crook, J. H. 1964. The evolution of social organization and visual communication in the weaver birds (Ploceinae). Behaviour Supplements 10:1-178.

Furrer, R. K. 1974. Nest site stereotypy and optimal breeding strategy in a population of Brewer's Blackbirds (Euphagus cyanocephalus). Aku-Fotodruck, Zurich.

———. 1975. Breeding success and nest site stereotypy in a population of Brewer's Blackbirds (Euphagus cyanocephalus). Oecologia 20:339-50.

Herrick, F. H. 1935. Wild Birds at Home. Appleton-Century Co., New York.

Holcomb, L. G., and G. Twiest. 1968. Ecological factors affecting nest building in Red-winged Blackbirds. Bird-Banding 39:14-22.

Muir, John. 1894. The water ouzel. In: The Mountains of California, pp. 276-99.

Orians, G. H., M. L. Erckmann, and J. C. Schultz. 1977. Nesting and other habits of the Bolivian Blackbird (Oreopsar bolivianus). Condor 79:250-55.

Schaefer, V. H. 1980. Geographic variation in the insulative qualities of nests of the northern oriole. Wilson Bulletin 92:466-74.

Skutch, A. F. 1976. Parent Birds and Their Young. University of Texas Press, Austin.

Wiley, R. H., and M. S. Wiley. 1980. Spacing and timing in the nesting ecology of a tropical blackbird: comparison of populations in different environments. Ecological Monographs 50:153-178.

7. Parasitism

Darley, J. A. 1982. Territoriality and mating behavior of the male brown-headed cowbird. Condor 84:15-21.

Elliott, P. L. 1980. Evolution of promiscuity in the brown-headed cowbird. Condor 82:138-41.

Friedmann, H. 1929. The Cowbirds: A Study in the Biology of Social Parasitism. C. C. Thomas, Springfield, IL.

———. 1963. Host relations of the parasitic cowbirds. United States National Museum Publication No. 233.

———. H., L. F. Kiff, and S. I. Rothstein. 1977. A further contribution to knowledge of the host relations of the parasitic cowbirds. Smithsonian Contributions to Zoology no. 235.

Hamilton, W. J., III, and G. H. Orians. 1965. Evolution of brood parasitism in altricial birds. Condor 67:361-82.

King, A. P., and M. J. West. 1977. Species identification in the North American cowbird: appropriate responses to abnormal song. Science 195:1002-4.

Lowther, P. E., and S. I. Rothstein. 1980. Head-down or "preening invitation" displays involving brown-headed cowbirds. Condor 82:459-60.

Mayfield, H. F. 1960. The Kirtland's warbler. Bulletin of the Cranbrook Institute of Science 40.

Payne, R. B. 1977. The ecology of brood parasitism in altricial birds. Annual Review of Ecology & Systematics 8:1-28.

Pearson, D. L. 1974. Use of abandoned cacique nests by nesting Troupials *(Icterus icterus):* precursor to parasitism? Wilson Bulletin 86:290-91.

Pinto, O. 1975. Sobre a apropriaçao dos ninhos de *Pseudoseisura cristata* (Spix) por *Icterus icterus jamacaii* (Gmelin). Papeis Avulsas de Zoologia, Sao Paulo 29:35-36.

Post, W., and J. W. Wiley. 1976. The yellow-shouldered blackbird—present and future. American Birds 30:13-20.

Rothstein, S. I. 1975. Evolutionary rates and host defenses against avian brood parasitism. American Naturalist 109:161-76.

———. 1975. Mechanisms of avian egg-recognition: do birds know their own eggs? Animal Behavior 23:268-78.

———. 1977. The preening invitation or head-down display of cowbirds. I. Evidence for intraspecific occurrence. Condor 79:13-23.

———. 1980. The preening invitation or head-down display of parasitic cowbirds. II. Experimental analysis and evidence for behavioral mimicry. Behaviour 75:148-82.

Selander, R. K., and C. J. LaRue, Jr. 1961. Interspecific preening invitation display of parasitic cowbirds. Auk 78:473-504.

Smith, N. G. 1968. The advantage of being parasitized. Nature 219:690-94.

Yom-Tov, Y. 1980. Intraspecific nest parasitism in birds. Biological Reviews 55:93-108.

8. The Roles of the Sexes

Lenington, S. 1980. Female choice and polygyny in red-winged blackbirds. Animal Behavior 28:347-61.

Orians, G. H. 1969. On the evolution of mating systems in birds and mammals. American Naturalist 103:589-603.

Picman, Y. 1980. Impact of marsh wrens on reproductive strategy of red-winged blackbirds. Canadian Journal of Zoology 58:337-50.

Searcy, W. A. 1979a. Female choice of mates: a general model for birds and its application to red-winged blackbirds *(Agelaius phoeniceus).* American Naturalist 114:77-100.

———. 1979b. Male characteristics and pairing success in red-winged blackbirds. Auk 96:353-363.

Selander, R. K. 1972. Sexual selection and dimorphism in birds. In: B. Campbell (ed.). *Sexual Selection and the Descent of Man.* Aldine Publishing Co., Chicago, pp. 180-230.

Trivers, R. L. 1972. Parental investment and sexual selection. In: B. Campbell (ed.), *Sexual Selection and the Descent of Man, 1871-1971.* Aldine Publishing Co., Chicago.

Wittenberger, J. F. 1980. Feeding of secondary nestlings by polygynous male bobolinks in Oregon. Wilson Bulletin 92:330-40.

9. On Communication

Baker, R. R., and G. A. Parker. 1979. The evolution of bird coloration. Philosophical Transactions of the Royal Society of London (B) 287:63-130.

Darwin, C. 1873. *The Expression of the Emotions in Man and Animals.*

Marler, P. 1961. The logical analysis of animal communication. Journal of Theoretical Biology 1:295-317.

Morris, L. 1975. Effect of blackened epaulets on the territorial behavior and breeding success of male red-winged blackbirds. Ohio Journal of Science 75:168-76.

Peek, F. W. 1972. An experimental study of the territorial function of vocal and visual display in the male red-winged blackbird *(Agelaius phoeniceus).* Animal Behavior 20:112-118.

Smith, D. G. 1972. The role of the epaulets in the red-winged blackbird *(Agelaius phoeniceus)* social system. Behaviour 41:251-68.

Smith, W. J. 1977. *The Behavior of Communicating: An Ethological Approach.* Harvard University Press, Cambridge, MA.

10. Black Is Conspicuous and Sometimes Warm

Burtt, E. H., Jr. 1981. The adaptiveness of animal colors. BioScience 31:723-29.

Hamilton, W. J., III. 1973. *Life's Color Code.* McGraw-Hill, New York.

Hardy, J. W., and R. W. Dickerman. 1965. Relationships between two forms of the red-winged blackbird in Mexico. The Living Bird 4:107-30.

Rowher, S., and P. W. Eward. 1981. The cost of dominance and advantage of subordination in a badge signaling system. Evolution 35:441-54.

Udvardy, M. D. F. 1972. Plumage cycles in adult North American birds. Aquila Serie Zoologia 13:56-60.

Walsberg, G. E., G. S. Campbell, and J. R. King. 1978. Animal coat color and radiative heat gain: a re-evaluation. Journal of Comparative Physiology 126:211-22.

11. Talk Is Cheap

Avery, M. and L. W. Oring. 1977. Song dialects in the bobolink *(Dolichonyx oryzivorous).* Condor 79:113-18.

Beletsky, L. D., S. Chao, and D. G. Smith. 1980. An investigation of song-based species recognition in the red-winged blackbird *(Agelaius phoeniceus).* Behaviour 73:189-203.

Falls, J. B., and J. R. Krebs. 1975. Sequences of songs in repertoires of western meadowlarks *(Sturnella neglecta).* Canadian Journal of Zoology 53:1165-78.

Falls, J. B., and L. J. Szijj. 1959. Reactions of eastern and western meadowlarks in Ontario to each other's vocalizations. Anatomical Record 134:560.

Feekes, F. 1977. Colony-specific song in *Cacicus cela* (Icteridae: Aves): the pass-word hypothesis. Ardea 65:197-202.

———. 1982. Song mimesis within colonies of *Cacicus cela* (Icteridae: Aves): a colonial password? Zeitschrift für Tierpsycholgie 58:119-52.

Ficken, R. W., and M. S. Ficken. 1963. The relationship between habitat density and the pitch of songs in Parulidae and other birds. American Zoologist 3:500.

Krebs, J. R. 1977. The significance of song repertoires: the Beau Geste hypothesis. Animal Behavior 25:474-78.

Lanyon, W. E. 1960. The ontogeny of vocalizations in birds. In: *Animal Sounds and Communication.* American Institute of Biological Sciences Publication No. 7.

Marler, P. 1959. Developments in the study of animal communication. In: Bell, P. R. (ed.), *Darwin's Biological work: Some Aspects Reconsidered.* Wiley, New York.

Orians, G. H. 1983. Notes on the behavior of the melodious blackbird. Condor 85:453-60.

Smith, D. G. 1979. Male singing ability and territory integrity in red-winged blackbirds *(Agelaius phoeniceus).* Behaviour 68:193-206.

———, and F. A. Reid. 1979. Roles of the song repertoire in red-winged blackbirds. Behavioral Ecology & Sociobiology 5:279-90.

Thorpe, W. H. 1961. *Bird Song: The Biology of Vocal Communication and Expression in Birds.* Cambridge University Press, Cambridge.

Wiley, R. H., and D. G. Richards. 1978. Physical constraints on acoustic communication in the atmosphere: implications for the evolution of animal vocalizations. Behavioral Ecology & Sociobiology 3:69-94.

Yasukawa, K. 1981. Song repertoires in the red-winged blackbird *(Agelaius phoeniceus):* a test of the Beau Geste hypothesis. Animal Behaviour 29:114-25.

12. Of People and Blackbirds

Bond, J. 1974. *Birds of the West Indies.* Collins, London.

Devitt, O. E. 1964. An extension of the breeding range of Brewer's blackbird in Ontario. *Canadian Field Naturalist* 78:42-46.

Friedmann, H. 1955. The honey-guides. United States National Museum Bulletin 208.

Haffer, J. 1969. Speciation in Amazonian forest birds. *Science* 165:131-37.

———. 1974. Avian speciation in tropical South America. Publication of the Nuttall Ornithological Club, no. 14.

Myers, N. 1979. *The Sinking Ark.* Pergamon Press, Oxford and New York.

National Academy of Sciences. 1980. *Conversion of Tropical Moist Forests.* A report prepared by Norman Myers for the Committee on Research Priorities in Tropical Biology of the National Research Council.

Soulé, M. E., and B. A. Wilcox. 1980. *Conservation Biology: An Evolutionary-Ecological Perspective.* Sinauer Associates, Sunderland, MA.

Stepney, P. H. R., and D. M. Power. 1973. Analysis of eastern breeding expansion of Brewer's blackbird plus general aspects of avian expansions. Wilson Bulletin 85:452-64.

Taverner, P. A. 1934. *The Birds of Canada.* Canada Department of Mines, Canada National Museum Bulletin 72.

Winterhalter, B., and E. A. Smith (eds.). 1981. *Hunter-Gatherer Foraging Strategies: Ethnographic and Archaeological Analyses.* University of Chicago Press, Chicago.

Tricolored blackbird

Index